Studies of Brain Function, Vol. 10

Studies of Brain Function

Volumes already published in the series:

Ulrich Bässler

Neural Basis
of Elementary Behavior
in Stick Insects

With 124 Figures

Springer-Verlag
Berlin Heidelberg New York 1983

Professor Dr. Ulrich Bässler
Fachbereich Biologie der Universität
Postfach 3049
6750 Kaiserslautern, FRG

Translated from the German by:
Dr. Camilla Mok Zack Strausfeld
Plöck 52, 6900 Heidelberg, FRG

ISBN 3-540-11918-3 Springer-Verlag Berlin Heidelberg New York
ISBN 0-387-11918-3 Springer-Verlag New York Heidelberg Berlin

Library of Congress Cataloging in Publication Data. Bässler, Ulrich, 1932–
Neural basis of elementary behavior in stick insects. (Studies of brain function ; v. 10) Includes index. 1. *Carausius morosus*—Behavior. 2. *Carausius
morosus*—Physiology. 3. Neurophysiology. 4. Insects—Behavior. 5. Insects
—Physiology. I. Title. II. Series. Q1508.B38B37 1983   595.7'24   82-19529

Offsetprinting and binding: Konrad Triltsch, Graphischer Betrieb, Würzburg
2131/3130-543210

Preface

This monograph represents the current status of neuroethological research on the diurnal behavior of the stick insect, *Carausius morosus*. The growing profusion of interrelated studies, many of which are published only in German, makes an overview of this field increasingly difficult. Many stick insect results contribute to general problems like control of catalepsy, control of walking, program-dependent reactions and control of joint position. For this reason I decided to compile and synthesize the results that are presently available even though the analyses are far from concluded. In addition to both published and unpublished results of the group in Kaiserslautern (Bässler, Cruse, Ebner, Graham, Pflüger, Storrer, as well as doctoral and masters students), I have drawn upon the literature which had appeared as of summer 1981. This includes above all the work of Godden and of Wendler and his colleagues in Cologne. A summary of the anatomical and physiological background, necessary for an understanding of these investigations, is provided in an appendix (Chap. 6). Methodological details must be obtained from the original publications. Figures for which no source is given are from my own studies.

I intend to update this monograph on an annual basis. Requests for these supplements should be directed to me in Kaiserslautern.

I would like to express my appreciation to all members of the group in Kaiserslautern for their constructive discussions, their unflagging cooperation, and their permission to include hitherto unpublished results. In addition, I would like to thank Prof. E. Florey, Dr. H.-J. Pflüger, Prof. G. Wendler, and Dr. C. Walther for critical review of the manuscript. Many of the Kaiserslautern studies were supported by grants from the Deutsche Forschungsgemeinschaft; and some of the earlier studies, from the Stiftung Volkswagenwerk.

Kaiserslautern, December 1982   Ulrich Bässler

Contents

1 Introduction

1.1 Statement of the Problem

Ethology and neurophysiology were once two widely separated areas of biology. Although behavior is obviously generated by the nervous system and its associated sense organs and muscles, its early study was primarily concerned with formulating an internally consistent system of principles to explain behavior without actually investigating the neurophysiological basis of these principles. Such an approach is both legitimate and practical, and every science dealing with problems of a certain complexity uses a similar approach. In chemistry, for example, it was and still is much more reasonable to formulate and work within an intrinsic system of principles than to wait for such laws to be deduced from nuclear physics.

Clearly it is more practicable to find intrinsic principles at a higher level of complexity (e.g., behavior), but more advanced stages of a science should also try to relate such principles to those at the next lower level of complexity (e.g., neurophysiology). In the investigation of neural mechanisms of behavior this conviction is embodied in the attempt to demonstrate that a few simple behaviors are characteristic properties of particular neuronal networks and their associated sense organs and effectors. Such research involves not so much a description of all neuronal activity during a behavior, but rather an investigation of the determinants of this activity. Since causal analysis is generally attributed to the realm of physiology, the study of neural mechanisms of behavior can be termed behavioral physiology, and Hoyle (1975) has also called it neuroethology.

The ideal experimental animal for the study of behavioral physiology should have a simply organized nervous system with few enough neurons that one can entertain the hope of tracing a particular behavior to the neuronal level. In addition, the animal should show only simple behaviors so that the analysis is within naturally given constraints. Such behaviors should be widespread within the animal kingdom so that the conclusions reached have a general applicability, perhaps providing insight into the neural basis of behavior of animals that are more difficult to investigate at the neuronal level for reasons such as the greater number of neurons involved.

1.2 The Experimental Animal and Its Behavior

The stick insect *Carausius morosus* meets all these requirements. *Carausius* is a relatively large insect. If it is too small for a particular experiment, there are many very large related species available (amongst them the largest known insects). Its anatomy is clearly arranged, and its nervous system is readily accessible, making it possible to obtain both intra- and extracellular electrophysiological recordings from behaving animals. The behavioral repertoire of the stick insect is simple, can be externally controlled to a great extent, and has components which are widely distributed in the animal kingdom, including mammals. Stick insect behavior so obviously subserves the purpose of twig mimesis that conclusions can also be drawn about the ecological significance of behavioral components as well as the coevolution of body shape and behavior.

This book deals only with the behaviors that adult *Carausius* exhibits under daylight conditions and in which the legs participate. This excludes all nocturnal behavior (these animals are active mainly at night during which they drink and eat), egg laying (e.g., Thomas 1979), and color change (e.g., Bückmann 1979). The sparse behavioral repertoire that remains consists of the following components:

1. *Death-feigning* (often called thanatosis) in which a posture is assumed which especially emphasizes the stick-like body shape (Figs. 2.1 and 2.3).
2. *No spontaneous activity* under daylight conditions (Sect. 2.2).
3. *Catalepsy:* If one forces a leg into a particular position and releases it, it returns so slowly to its original position that it appears to an observer to be immobile (Fig. 2.4). This behavior is called catalepsy or *flexibilitas cerea.* Similar behavior is found in many animals, including humans and other mammals (where it is considered pathological).
4. *Rocking* consists of rhythmic movements primarily in a direction perpendicular to the longitudinal axis of the body. They often appear after the animal has been slightly disturbed (Sect. 2.5.1). There are certain parallels between rocking movements and tremor phenomena in man and other mammals.
5. *Active movements* brought about by severe disturbance are often irregular in restrained animals.
6. *Walking:* unrestrained animals often walk after being disturbed. Walking movements are carried out basically in the same way as in all animals with this mode of locomotion.
7. *Orientation:* During walking the animals orient to light, gravity, and/or internal cues.

Each of these behaviors is described in detail in the following chapters. Emphasis is placed on the more generally distributed behaviors (3), (4), and (6). In Chapter 2 the first behaviors discussed are those involved in the camouflage of the animal (the so-called twig mimesis) since these are the best understood.

1.3 Experimental Strategy

Behavioral physiology bridges two fields of varying complexity, ethology and neurophysiology. The history of science has shown that such a bridge is in general more easily constructed starting from the more complex of two levels. Most investigators of *Carausius* behavior have planned their strategy accordingly. This strategy for the investigation of a single behavior can be divided into three stages:

1. Quantitative description of the behavior. This can be a description of the movement itself, of muscle forces, or of neuronal activity during the behavior.
2. Relating the described behavior to the activity of one or a few unambiguously defined systems. In other words, one tries to demonstrate that the behavior in question is a characteristic of a specified system. For example, it can be shown that catalepsy in the femur-tibia joint is a characteristic of the control system governing this joint (see Sect. 2.4.2). This system, although not fully understood, is clearly defined as the feedback loop from the femoral chordotonal organ to the extensor and flexor muscles of the knee joint, including the neuronal structures which determine the reference position of the joint and the gain of the feedback loop. In addition to electrophysiological and behavioral studies, systems theory (cybernetics) is indispensable for formally describing the characteristics of a system.
3. Elucidation of the neuronal base of the system, which has already been defined functionally, primarily by electrophysiological methods.

As will be seen the study of catalepsy is presently in Stage 3 (Sect. 2.8), Stage 2 having been completed (Sects. 2.4.2 and 2.5.3). The analysis of rocking and walking is still in Stage 2.

2 Behavioral Components of Twig Mimesis – Experiments on the Femur-Tibia Joint

2.1 Twig Mimesis and Its Components

Even for a trained human observer, it is almost impossible to detect a stick insect amidst a tangle of branches. As in many other phasmids twig mimesis is the most important protective mechanism of the stick insect. Except for the ability to autotomize its legs, *Carausius* does not possess any of the other defense mechanisms found in the phasmids (like ant mimicry and defense clap in *Extatosoma*, sudden appearance of color signals in *Orxines*, or defense glands in *Anisomorpha*, Clark 1974; Beier 1957, 1968). Thus, twig mimesis is apparently the decisive factor for the entire biology of the stick insect.

Fig. 2.1. Stick insect hanging from a branch

Twig mimesis consists of two components, appearance and behavior. The body shape is very twig-like (see Fig. 2.1), and during the day when the animal can be seen, its passive behavior further emphasizes its resemblance to a twig.

The twig-like *form of the body* is mainly due to (1) the elongation of the entire body, especially the meso- and metathoracic segments, (2) the great mobility of the subcoxal joints which allows the legs to be held parallel to the longitudinal axis of the body, and (3) indentations in the prothoracic legs which enclose the head so that when the legs are extended forward, a direct optical elongation of the body is achieved (for details, see Sect. 6.1).

Fig. 2.2. The stick insect *Carausius morosus*. Its length excluding the antenna is 7–8 cm

The *behavioral components* pertaining to twig mimesis are discussed in the following chapters and can be summarized as follows:

1. The stick insect is not spontaneously active in daylight (see Sect. 2.2).

2. During the day the stick insect typically assumes a twig-like posture (stick posture, see Sect. 2.3).

3. Due to the mechanism of catalepsy the return of the legs to their original position after passive displacement or brief active movements is so slow that the leg appears to be immobile (see Sect. 2.4).

4. Rhythmical movements, mainly in the direction of the transverse axis of the animal, appear after disturbance. They appear to reduce the probability of the insect's being eaten (Rupprecht 1971) perhaps because these movements increase the resemblance to a twig swaying in the wind or because many predators do not regard objects that move perpendicularly to a pronounced longitudinal axis as prey (Ewert and Wietersheim 1974). Brief active movements and finally escape movements (walking away) appear only after a severe disturbance such as abdominal pinching (see Sect. 2.5).

5. Stick insects often release their hold on a branch and drop toward the ground after a vigorous disturbance. During their fall they usually assume the stick posture with their legs laid flat against their body. The tarsi are bent, forming small hooks with which the animal catches the lower branches. Such a fall quickly moves the animal out of reach of an attacker without forcing the stick insect to leave the branches which provide its food supply.

2.2 Control of the State of Activity

Light seems to be primarily responsible for the suppression of spontaneous activity. Constant bright light can suppress activity for so long that the animal eventually starves to death (Steiniger 1933; Kalmus 1938; Eidmann 1956; Godden and Goldsmith 1972). Green light is maximally effective, but red light at adequate brightness can also suppress activity (Godden and Goldsmith 1972). The lower the light intensity following a period of constant bright light, the more quickly the animal resumes activity. Although the eyes are the receptors that are primarily responsible for this behavior, bright light can still suppress activity even if the eyes are painted over or removed. The threshold for this suppression is, however, appreciably higher than when the eyes are intact. Since the stick insect does not possess ocelli, there must be another receptor system (perhaps a light sensitive part of the CNS or an epidermal light receptor) which is responsible for the response with a higher threshold (Godden and Gold-smith 1972; Zwenker 1973).

According to Kalmus (1938) and Eidmann (1956), stick insects that have been under a natural light–dark regime maintain their circadian rhythm in reduced form for at least a few days after being placed in constant dark. In contrast to animals with a natural light–dark cycle, animals in constant dark are active at all times of day although certain activity maxima corresponding to the dark phase of the normal cycle are still evident.

Under more standardized conditions Godden (1973) repeated these experiments and found that after the animal was exposed to constant dark, the original rhythm either completely disappeared or inverted, i.e., there was a phase shift of 12 h in the activity rhythm.

These findings suggest that the circadian rhythm plays only a subordinate role in the control of activity under normal light-dark conditions. The priority of suppression of activity by light is a guarantee that the insect remains motionless as long as it can be seen by predators.

Decerebrate Animals: It has long been known that stick insects without a brain (supraesophageal ganglion) are active at all times of day and are easily excitable (e.g., Steiniger 1933; Eidmann 1956). Graham (1979a) has precisely described the behavioral repertoire of such animals. To "decerebrate" with minimal blood loss, he cut only the connectives between the supra- and subesophageal ganglia. Such decerebrate animals show the following behaviors:

1. *Resting.* The animals stand quietly with normal body posture but never in the stick posture.
2. *Twitching.* These small movements of the body are caused by an abrupt increase in muscle tone. Myograms from pairs of antagonistic muscles, the protractor and retractor coxae and the flexor and extensor tibiae (Pflüger 1976), reveal a synchronous increase in activity in both muscles of a pair during a twitch (for location and innervation of the muscles see Sects. 6.1.2 and 6.3). Twitching occurs in a series of 3–8 bouts which increase in intensity and which are separated by decreasing intervals (from 4 s at the beginning to about 1 s towards the end of a series). The abdomen is raised above the body in a scorpion-like manner. During the last twitch the legs are thrown vigorously forward which normally results in a backward jump, followed by walking.
3. *Walking.* The step frequency starts high and decreases gradually. After a period of time the animal returns to the resting state.

An activity cycle usually begins with a series of twitches, followed by walking. Activity can spontaneously appear shortly after surgery. Later a light touch is enough to release as many as six activity cycles, interrupted by short pauses. Twitching is not restricted to decerebrate animals. Intact animals twitch occasionally but usually with a low intensity.

2.3 Thanatosis

Stick insects can assume several different postures in the daytime. At one extreme, the simple resting position, the femur-tibia joints are bent and the femurs stand away from the body (Figs. 2.2 and 2.3). At the other extreme, the stick posture, all the femur-tibia joints are fully extended with the forelegs stretched forwards and the middle and hindlegs lying flat against the body (middle legs forward and hindlegs to the rear). The transition between simple resting position and stick posture is continuous (e.g., an intermediate position would be with forelegs fully extended anteriorly and the other legs standing away from the body, see Figs. 2.1 and 2.3).

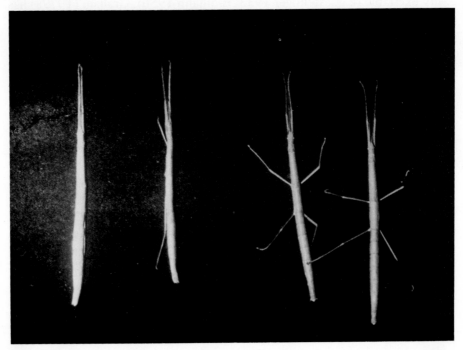

Fig. 2.3. Diverse resting positions with the stick posture on the *left* and the simple resting position on the *right*

In the stick posture animals often align themselves parallel to edges in the visual field (Jander and Volk-Heinrichs 1970). In its natural setting an animal lying parallel to a branch looks like a broken-off twig.

Distinguishing between different states of thanatosis by defining individual postures, as has been done in the past (e.g., Steiniger 1933), is rather unwieldy and arbitrary because of the gradual transition between the two extreme postures. A better alternative for characterizing thanatosis is to use a single variable that is minimal for one of the extremes and maximal for the other. Following the example of earlier authors this variable is termed the depth of thanatosis. The simple resting position corresponds to the minimum and the stick posture to the maximum depth.

The depth of thanatosis is here defined as the tendency for the femur-tibia joint to reach its fully extended position. This tendency can be measured quantitatively as the extension force exerted by the tibia at a femur-tibia angle of 170° (Bässler 1972c). The greater the force, the deeper the thanatosis. Electrophysiologically it can be expressed as the impulse frequency of the slow extensor tibiae motor neuron (SETi); the higher the frequency, the greater the depth. SETi is almost always spontaneously active (Godden 1974) and controls the position of the femur-tibia joint (see Sects. 2.8 and 6.3.3).

The use of these measures is supported by the following: (1) They can describe all thanatotic positions; the simple resting position with flexed femur-tibia joint gives a negative force reading and very low SETi activity. (2) Animals which can be made active by weak mechanical stimulation give negative or weakly positive force readings whereas animals which remain motionless after the same stimuli give force readings which are more strongly positive (Bässler 1972c). (3) At the beginning of a dark period the SETi activity falls gradually until the appearance of the first active movement (Godden 1974).

The inactive state can be terminated by various stimuli (heat, strong mechanical stimulation, turning off the lights) regardless of the depth of thanatosis (Godden 1972; Steiniger 1933).

Earlier publications on thanatosis which treated it as a state rather than a variable often proposed that it was a condition of decreased reflex excitability. It has since been found, however, that the force which opposes passive flexion of the femur-tibia joint (used as a quantitative measure for reflex excitability) is not negatively correlated with the above measure for depth of thanatosis (see also Sect. 2.6.3). The term thanatosis, as used here, thus differs in two ways from its earlier usage. It does not refer to a distinct position but to a range of positions, and the depth of thanatosis is not negatively correlated with reflex excitability.

If the legs are in the stick posture before the prothorax is cut off, they remain in this position after the operation. In isolated meso- or metathorax preparations the position of the legs relaxes after a time (Godden 1972). Apparently the meso- and metathorax, but not the prothorax, need nervous input from other segments for the maintenance of the stick posture.

After removal of the supraesophageal ganglion or cutting of the esophageal connectives, the animals walk spontaneously (see Sect. 2.2). Between walks they assume a simple resting position, but never, not even in the dark, do they assume the stick posture. Thus, for all segments information from the supraesophageal ganglion is essential for the triggering of the stick posture.

To an observer the femur-tibia joint of an inactive stick insect appears to be completely immobile. However, if the joint position is registered continuously over a longer period of time, several kinds of movements can be observed (Kittmann 1979): (1) "Fast" movements with amplitudes mostly under 5° occur in both directions. The movement duration is less than 5 s. Usually the legs return to their starting position within a few seconds. (2) "Slow" movements also have amplitudes mostly under 5° and durations between 20 and 150 s. They also occur in both directions but the legs usually do not return to their starting position. (3) Occasionally a rhythmic flexion and extension with a frequency of 0.1 Hz and an amplitude of 0.5°–3° can be observed for hours. (4) Twitching movements (some of which are statistical) with very small amplitudes are superimposed on the flexion and extension. (5) Very slow return movements occur after active (elicited by disturbance) or passive movements of larger amplitude (see Sect. 2.4). These experiments show that the resting position is not maintained by friction or elastic properties of the muscles but rather by the active use of particular muscles.

2.4 Catalepsy *(Flexibilitas Cerea)*

2.4.1 Description and Definition of Catalepsy

If a leg joint of an inactive stick insect is passively bent, held for a short time in this position, and then released, it appears to the casual observer to remain in this position (Fig. 2.4). In the earlier literature this behavior was attributed to the plasticity of the musculature combined with a strong decrease in reflex excitability (Schmidt 1913; Rabaud 1919; Steiniger 1933). This behavior was called catalepsy or *flexibilitas cerea* (waxen flexibility) because of its resemblance to the cataleptic (catatonic) state in humans and other mammals.

Quantitative investigations of this phenomenon have been carried out on the femur-tibia joint by Bässler (1972c) and Godden (1974). In these

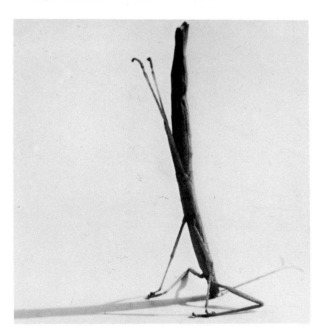

Fig. 2.4. Catalepsy in *Carausius*. The legs were manually forced into their positions and the insect was stood on its head. The animal "froze" in this unnatural position. (According to an idea from Schmidt 1913)

experiments the femur-tibia joint was bent from the extended position to a particular angle, held in this new position for a defined time interval, and then released (Fig. 2.5). It was found that immediately after release, the joint extends quickly about 10°–30° (fast phase) and then returns slowly to the starting position with ever-decreasing speed (slow phase). The velocity of the slow phase lies between 180°/min (half the angular velocity of the second hand of a clock) and 0.1°/min (which is even slower than the hour hand of a clock (0.5°/min). Thus the velocity of the slow phase can vary greatly.

If the leg is held in a new position for periods of time shorter than 2 s, the angle through which the leg moves in the fast phase depends on the holding time (the shorter the holding time, the greater the angle). Increasing the holding times beyond 2 s does not further decrease the angle significantly. If the above experiment is repeated several times consecutively, the return velocity in the slow phase decreases gradually with each repetition. This property and the variability of the data are probably the reason why no significant dependence of the return velocity on the holding time and bending velocity can be demonstrated (Godden 1974).

Since the leg does not remain in the imposed position during catalepsy, there is no basis for the assumption that the musculature possesses a high degree of plasticity. Although the original definition for catalepsy and *flexibilitas cerea,* which has gained wide acceptance in the literature, includes this interpretation, we continue to use the term catalepsy but only for describing and not interpreting this phenomenon.

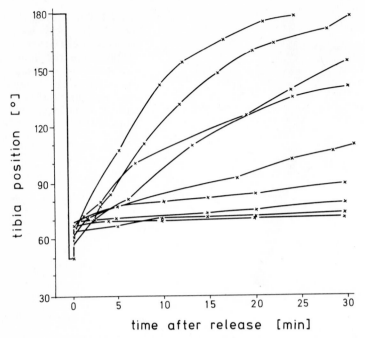

Fig. 2.5. Return movements of the tibia to the 180° starting position after being passively bent to 50° for 30 s. Nine randomly selected examples of intact legs

Catalepsy is particularly conspicuous when the femur-tibia joint is in the extended position (great depth of thanatosis). This suggests that thanatosis and catalepsy may only be different expressions of the same condition. Steiniger (1933) made this assumption and treated them both as catalepsy. It has been shown, however, that the depth of thanatosis and the return velocity are weakly but significantly correlated with each other, i.e., the deeper the thanatosis, the faster the tibia returns to the extended position (for quantitative measures of thanatosis see Sect. 2.3). A high return velocity means a smaller degree of catalepsy. This negative correlation between the depth of thanatosis and the degree of catalepsy (Bässler 1972c) makes it inadmissable to combine the two conditions under the single concept of catalepsy.

This conclusion is also supported by another finding. The return movements following rapid active movements of greater amplitude (released by disturbance, see Sect. 3.2) are indistinguishable from return movements after passive movements of inactive animals (Bässler 1973; Kittmann 1979; see also Sect. 3.2.1) and must therefore also be regarded as catalepsy.

In accordance with these considerations, catalepsy is defined here as the slow return movement after a large amplitude movement, whether it be passive or active. No causal interpretation is included in the definition.

2.4.2   Catalepsy in the Femur-Tibia Joint as a Characteristic of a Feedback Control Loop

Let us first consider catalepsy after a passive movement. If the position of a joint is stabilized by a feedback control loop, this loop will produce a resistance reflex to any passive movement of the joint. Movement is a necessary prelude to the onset of catalepsy. If a feedback control loop is functioning in this joint during catalepsy, this loop must participate in the manifestation of catalepsy. The following experiments demonstrate that catalepsy in the femur-tibia joint is a characteristic of the feedback control loop which governs this joint. The femur-tibia joint was selected for these investigations because it is anatomically well-suited to such studies. The position of this joint is measured primarily by the femoral chordotonal organ (see Sect. 6.2.3 and Fig. 2.6).

A very simple experiment demonstrates that the position of the femur-tibia joint is stabilized by a feedback control loop. Normally even a relatively sizable force exerted on the tibia displaces it only slightly from its original position. However, if the receptor apodeme of the femoral chordotonal organ is cut (see Sect. 6.2.3 and Fig. 2.6), a slight pressure can produce a large deflection of the tibia (Bässler 1965). Apparently this control loop continues to function during catalepsy since there is a distinct resistance during the beginning of a fast deflection.

The quantitative study of a feedback control loop is facilitated by opening the loop. In this system it can be accomplished in the following way. An incision is made in the femur at the position shown in Fig. 2.6, and the receptor apodeme is severed. A clamp, which is attached to the end of the receptor apodeme leading to the chordotonal organ, can be used to move the receptor apodeme and give the chordotonal organ a defined stimulus. Pulling the receptor apodeme (= stretching the chordotonal organ as occurs when the joint is flexed in the closed loop system) elicits an extension of the joint. Pushing the receptor apodeme in the

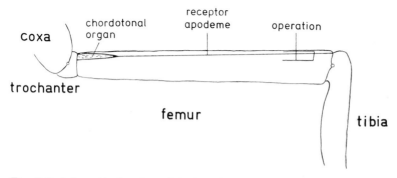

Fig. 2.6. Schematic drawing of the leg of a stick insect. The rotational axes are indicated by *small circles* for the coxa-trochanter and the femur-tibia joint

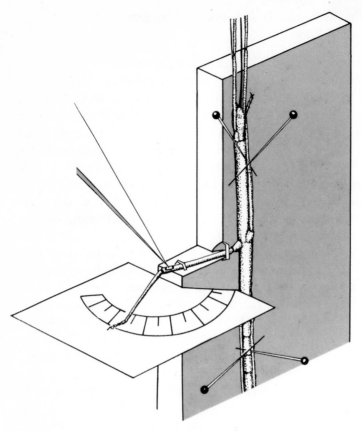

Fig. 2.7. Part of the experimental set-up for quantitative measurements on the open loop system

direction of the body (= releasing the chordotonal organ) results in a flexion of the femur-tibia joint. The movement of the tibia that is elicited by a particular stimulus can be recorded.

Figure 2.7 illustrates part of the experimental set-up. The animal is fastened to a cork with one leg held perpendicular to the side of the body. The angle formed by the femur and tibia of this leg is monitored (symbolized in the illustration by a protractor, for details see Bässler 1972b; Ebner and Bässler 1978; Kittmann 1979). Since in daylight spontaneous movements of large amplitude have never been observed, one can be certain that the recorded movements of the tibia are in response to stimulation of the chordotonal organ.

Figure 2.8 illustrates a typical response to a stepwise movement of the receptor apodeme. The response to such a stimulus is called a *step response*. The tibia responds sharply to the onset of the stimulus but slowly returns almost to its starting position if the stimulus is maintained. This

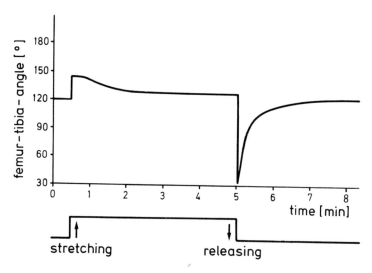

Fig. 2.8. Schematicized response of the tibia *(upper curve)* to stepwise stimulation of the femoral chordotonal organ *(lower curve)*

remaining offset from the starting position could be due solely to the mechanical friction of the joint and the musculature. In order to test this possibility the forces produced by the extensor and flexor tibiae, the two muscles which move the femur-tibia joint (see Sect. 6.1.2) were recorded separately. The force produced by the extensor tibiae muscle in response to a stepwise stretching of the chordotonal organ rises quickly and does not completely return to the starting value even after 10 min. The flexor tibiae muscle shows a similar response. These remaining forces are very small (under 1 mN compared to maximum forces of 80 mN for the extensor and 350 mN for the flexor, unpublished). Thus the system responds primarily to movement. This phasic behavior with a long decay time is referred to as the dynamic portion. There is also a slight tonic constituent, referred to as the static portion. The remaining offset from the starting position is mainly produced by this static portion and not by mechanical friction. In the control system the dynamic portion predominates and effectively brakes movements. It is, however, unable to compensate for maintained deflections from a reference position.

The amount of extension elicited by a stepwise stretching of the chordotonal organ is always less than the amount of flexion elicited by releasing it the same distance. The amplitude of both extension and flexion is approximately proportional to the logarithm of the stimulus amplitude. The more the chordotonal organ is prestretched before stimulus onset, the greater the amplitude of the movement in both extension and flexion (Figs. 2.9 and 2.10; Bässler 1965).

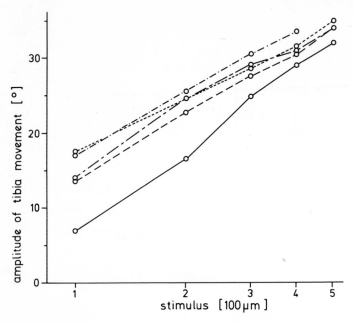

Fig. 2.9. Amplitude of tibia movement (extension) in response to stretching the chordotonal organ as a function of stimulus amplitude. Each *curve* represents the response following a particular degree of prestretching of the chordotonal organ and each *data point* is the mean of 16 measurements. The amount of prestretching for —— = 0 μm; ——— = 100 μm; —·— = 200 μm; - - - - = 300 μm; and —·— = 400 μm

After a stimulus the tibia returns almost to its starting point. This starting position, which can be maintained for a long time, varies widely from one animal to another, and even the same individual can show different starting positions from one day to the next (usually varying between 90° and 180°). The feedback control loop adapts almost completely and, therefore, is unable to stabilize a predetermined reference position over an extended period of time in the face of external influences. It can only brake movements. For this reason the starting position cannot be controlled by feeding a reference input into the control loop. Information about the starting position more probably reaches the motor neurons via some pathway other than the feedback loop (Bässler 1967).

The reaction time, i.e., the time between stimulus onset and the beginning of movement lies between 60 and 200 ms for stretching and 20–60 ms for releasing the chordotonal organ.

The return movement of the tibia during maintained stimulation (time of adaptation) approximates an exponential function, which can be characterized by its half-life. The half-life of the time course of adaptation (half-life of decline), lies between 10 and 240 s (with a median value of

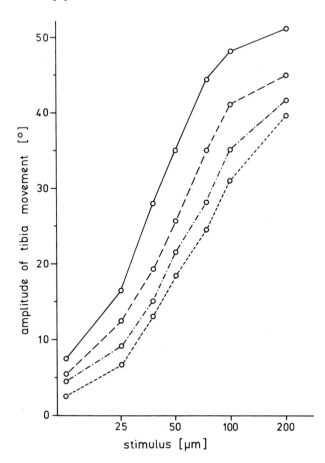

Fig. 2.10. Amplitude of tibia movement (flexion) in response to releasing the chordotonal organ as a function of stimulus amplitude. Each *curve* represents the response following a particular degree of prestretching of the chordotonal organ and each *data point* is the mean of 12 measurements. The amount of prestretching for ——— = 800 μm; — — — — = 600 μm; — · — · — = 400 μm; and · · · · · = 200 μm

approx. 50 s) for stretching and between 3 and 60 s (median value, 12 s) for releasing the chordotonal organ (Bässler 1972a).

The very slow joint movement during the slow phase of catalepsy can be more closely simulated by a ramp-shaped (or triangular) stimulus with low, constant velocity than by a step stimulus. Figure 2.11 shows the results of an experiment using ramp stimuli. The position of the tibia is given as a function of the position of the receptor apodeme during a release and a stretch. The tibia position corresponding to the same receptor apodeme position differs markedly for stretch and release. Such a difference can be produced only by the velocity-dependent dynamic portion of the system and not by the static portion. The static portion cannot produce any difference because it is by definition only dependent on receptor apodeme position. At stimulus velocities of 5 μm min^{-1} and higher there is always a difference between the response to stretch and to release (albeit not always as striking as in Fig. 2.11). At a velocity of 2 μm min^{-1} the

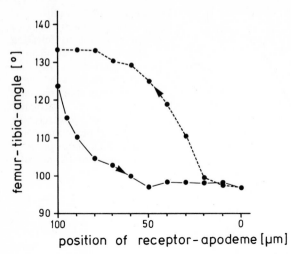

Fig. 2.11. Tibia position as a function of the receptor apodeme position during a release (———) and a stretch (– – – –) at a constant velocity of 5 μm min^{-1}. A receptor apodeme position of zero corresponds to a prestretch of 300 μm

difference, if it exists at all, is usually quite small. Thus, the threshold for the dynamic portion of the system seems to lie within this range (Bässler 1972a). The difference between the response to stretch and to release is a measure for the relative contribution of the dynamic portion of the system. This difference and consequently the contribution of the dynamic portion increase with increasing stimulus velocity. That the dynamic portion is dependent on stimulus velocity is also shown by Figs. 2.30 and 2.43.

The response of one animal to a particular stimulus can vary considerably (see Fig. 2.12) and is always strong after the animal has been disturbed. Response intensity is highest directly after an active movement elicited by a disturbance and declines gradually if the animal is subsequently left undisturbed. It is weakest directly after a spontaneous active movement released by exposing the animal to darkness after about 24 h in constant light (Bässler 1974).

Catalepsy of the femur-tibia joint can be regarded as a characteristic of the system that controls the position of the femur-tibia joint. Let us assume that the angle between the femur and the tibia has a starting position of 180°. Bending of the joint (to 50°, for example) meets with an opposing force, which is made up of two components. Component 1 is the dynamic portion of the control system plus a small force produced by the extensor muscle due to its viscosity (see Sect. 6.1.2). The magnitude of this component increases with increasing velocity of tibia movement and is zero when the movement stops. Component 2 consists of the static portion of the control system plus a second portion resulting from information about the starting position. As has already been discussed, this

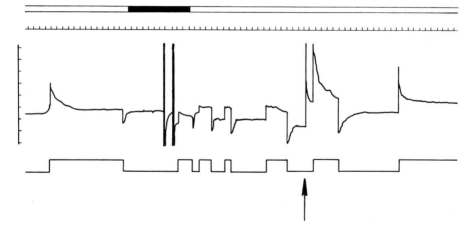

Fig. 2.12. A recording of the forces opposing a passive flexion and extension of the femur-tibia joint of an intact animal. *First row* light-dark cycle; *second row* time in minutes; *third row* force in mN with flexing force shown as an upward deflection and extending force, as a downward deflection; *fourth row* angle between femur and tibia with the *upper line* equal to 90° and the *lower* to 80°. *Arrow* denotes when the animal was touched on the abdomen. The animal was kept 28 h in the light before the beginning of the experiment. Note the spontaneous active movements during the dark cycle *(black bar)*

information about the starting position does not reach the motor neurons via the feedback loop. When the joint is passively bent, the extensor produces a small opposing force due to its elastic properties. Component 2 is practically independent of the velocity of tibia movement.

If the tibia is held in the 50° position for some time, component 1 decreases to zero, but component 2, which does not adapt, still acts on the tibia to bring it back to its starting position. Thus, after being released, the tibia will move toward the starting position, and the leg will straighten out. This extension of the leg produces an opposing tendency in the dynamic portion of the control system to stop the movement. Since this tendency first comes into effect after expiration of the reaction time, the tibia should move quickly at first and then abruptly slow down. This is precisely the behavior that has been observed (see Fig. 2.5).

The amplitude of the fast phase of the return movement will be greater, the higher the recoil force at the moment of release. If the holding time is short, the dynamic portion of the control system will not yet have decreased to zero. Thus, large amplitudes of the fast phase of the return movement are to be expected for short holding times and have actually been observed (see Sect. 2.4.1).

At the end of the fast phase of the return movement, the tibia comes under the influence of two opposing tendencies. One produced by component 2 moves the tibia towards the starting position. The other produced by component 1 acts as a brake on this movement. Component 2 is independent of velocity, but component 1 increases with increasing velocity and is zero when the velocity is zero. There will be a certain velocity where both components have equal value but opposite sign. This velocity of counterbalance will be the velocity of backwards movement, because at higher velocities component 1 exceeds component 2 and the movement is slowed down. At lower velocities component 2 is the higher one and the movement accelerates. As component 1 responds to very low velocities, the speed of backward movement will be very low.

Component 2 is a function of the displacement of the tibia from the starting position and decreases as it is approached. Component 1 is only slightly dependent on position (see Fig. 2.10) and decreases only minimally as the tibia moves toward its starting position. Thus, the return velocity will gradually decrease because component 1 will counterbalance component 2 at decreasing velocities as the starting position is approached (see Fig. 2.5). The whole system behaves like the model of Fig. 2.13. Component 1 is generated by a receptor apodeme displacement greater than approximately 2 μm min^{-1}. In the closed loop system this displacement corresponds to a joint movement of $0.2-0.3°$ min^{-1}. This should represent the lower limit of the return velocity because below that speed component 1 is zero. This velocity corresponds well to actual observations. All data available up to now support the hypothesis that catalepsy is a characteristic property of the closed-loop control system.

This explanation of catalepsy has also been confirmed by the following experiments:

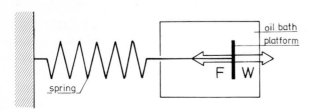

Fig. 2.13. Mechanical model of the slow phase of catalepsy. A spring pulls a platform from the right to the left side of an oil bath. Two forces act on the platform: the force produced by the spring (F) which is independent of the velocity of the platform movement (component 2) and the hydrodynamic force of resistance (W) which is zero when the velocity is zero and increases with increasing velocity (component 1). At the beginning, when the platform at the right side of the oil bath is released, F is larger than W. The resulting force accelerates the platform to that velocity where W counterbalances F. F decreases with movement toward the left. Therefore the velocity of counterbalance will also decrease

1. When the receptor apodeme is cut, the tibia returns relatively quickly to its starting position with no sudden decrease in return velocity (Bässler 1972c; Godden 1974).

2. The higher the gain of the control system (definition: Sect. 2.5.3) when measured just before or after a return movement in the intact system, the lower the velocity of the slow phase of the return movement. This agrees with the hypothesis because the higher the gain, the stronger the "braking" effect of component 1, which is now able to counterbalance component 2 at lower velocities (Bässler 1972c).

3. The deeper the thanatosis (high component 2), the faster the return movement (see Sect. 2.3). This correlation is not as strong as the one between gain and return movement velocity which is consistent with the weak correlation between gain and depth of thanatosis.

4. The control system is strongly asymmetric. After a passive extension component 2 is larger (the flexor is considerably stronger than the extensor) and component 1 is smaller (the control system responds less to stretching of the chordotonal organ) than after a passive flexion. In addition component 1 has a longer reaction time. After a fairly large passive extension, one would then expect the fast phase of a return movement to have a greater amplitude and the slow phase to have a higher velocity. Since *Carausius* very rarely assumes a femur-tibia angle less than 90°, it has been possible to passively extend the tibia over a large angle in only a few cases. These few cases, however, show precisely the predicted results (unpublished).

5. The motor output to the extensor and flexor muscles of the tibia changes in the manner required by the hypothesis (Godden 1974).

In accordance with these considerations, catalepsy can be regarded as a characteristic of the system which controls the femur-tibia joint. This concludes stage 2 of the strategy for the investigation of catalepsy in the femur-tibia joint (see Sect. 1.3).

2.5 Rocking

2.5.1 Description of Rocking

Rocking consists of rhythmic movements usually in the direction of the transverse axis of the body. They occasionally occur in the direction of other body axes and of combinations of axes, but only rocking movements in the direction of the transverse axis have been investigated in detail. Rocking is a very conspicuous behavior and has often been described

(Meissner 1909; Steiniger 1936; Beier 1968; Stabler 1976). Quantitative studies on rocking have been carried out by Rupprecht (1971) and in greater detail by Pflüger (1976, 1977).

In this behavior the stick insect moves its body rhythmically from side to side keeping its longitudinal axis parallel to itself. All the legs on one side of the body move in synchrony as long as they are not in the stick posture or raised off the ground. The angles between the coxa-trochanter, femur-tibia, and tibia-tarsus joints change during rocking (Fig. 2.14). The movement has an amplitude between 0.1 and 20 mm and a frequency between 1.2 and 5.6 Hz with a pronounced peak in distribution between 2 and 3 Hz.

Rocking is exhibited usually by resting animals which have been disturbed by external stimuli (air puffs, touch, shadows, warming, moistening, etc.). It can, however, also be observed, superimposed on walking move-

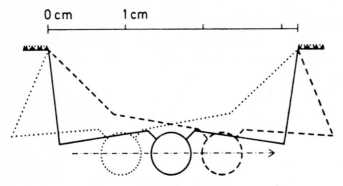

Fig. 2.14. Schematic cross-section through a thoracic segment of a hanging animal. Rocking parallel to the transverse body axis. (Pflüger 1976)

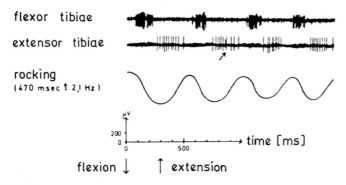

Fig. 2.15. Electrical activity in the flexor tibiae (myogram) and the extensor tibiae (nerve F2 recording) during rocking. The active unit in F2 is the SETi neuron. A single spike from the inhibitor is marked by an *arrow*. (Pflüger 1976)

ments. Electrophysiological recordings from unrestrained animals show that the activity of the flexor and extensor tibiae muscles in one leg alternate rhythmically during rocking. The extensor rhythm (Fig. 2.15) is maintained primarily by the SETi neurons (see Sect. 6.3.3). In the flexor only units with small muscle potentials are active. The common inhibitor neuron is rarely active (Pflüger 1976, 1977).

2.5.2 Possible Origins of Rocking Movements

Since the femur-tibia joint is one of the joints which is moved during rocking, the control loop of this joint must be involved in the movements. The gain of the femur-tibia control loop is especially high when the animal has been disturbed, i.e., when rocking occurs. Control loops with high gain often have very small phase reserves, i.e., they are not far from their point of instability. Thus, one possible explanation for rocking could be that a disturbance raises the gain of the femur-tibia control loop and other similar control loops so that they become unstable. In the following sections it is shown that although the femur-tibia control loop can in fact function very near the edge of instability, rocking movements are generated by a superimposed oscillator. The femur-tibia control loops only interfere with these imposed oscillations because of the resonance properties of the loops.

2.5.3 Frequency Response of the Open Femur-Tibia Control Loop

When the receptor apodeme is moved sinusoidally back and forth at different amplitudes and frequencies in the open loop configuration (see Sect. 2.4.2), movements of the tibia are produced which as a whole comprise the frequency response of the system. Figure 2.16 shows the response to sinusoidal stimulation at different frequencies with a stimulus amplitude of 100 μm. The amplitude of the tibia movement decreases rapidly with increasing frequency (Fig. 2.16, lower diagram). Such curves can be used to determine the phase shift between stimulus and response as well as the amplitude of the movement.

A phase shift can be unequivocally determined only when both curves have the same form (e.g., two sinusoidal curves). Since this is not the case here, we defined phase shift as the time interval between the onset of receptor apodeme movement in one direction and the onset of tibia movement in the corresponding direction. The amount of phase shift depends to some extent on the definition of phase shift. As other definitions are

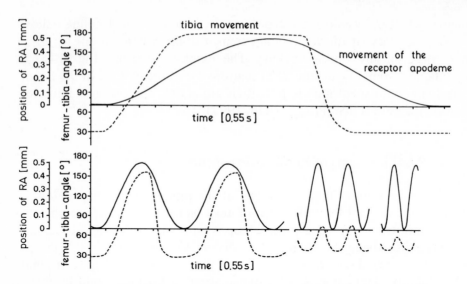

Fig. 2.16. Tibia movement in response to sinusoidal movement of the receptor apodeme at different frequencies

also conceivable, one must exercise caution when making inferences from such measurements. Phase shifts are generally not given in absolute time but in degrees (one full oscillation = 360°). Positive phase shift values indicate that the tibia movement occurs earlier than the corresponding receptor apodeme movement.

To facilitate later considerations on the stability of the system, the amplitude of the tibia movement must be given in the same dimension as the stimulus amplitude. This was accomplished by determining how much a particular movement of the tibia moves the receptor apodeme in the closed loop system. It was found that a movement of the tibia in its middle range of 13° resulted in a receptor apodeme movement of 100 μm. Thus, if a 100 μm stimulus in the open system elicits a tibia movement of 13°, the system has a gain of one (input amplitude = output amplitude). The gain will be two if the above stimulus produces a 26° tibia movement.

A Bode plot can be used to describe the dependence of gain (response amplitude as a multiple of the stimulus amplitude) and phase shift on frequency. Figure 2.17 shows the Bode plot for a stimulus amplitude of 100 μm (Bässler 1972b). The corresponding plots for other stimulus amplitudes are similar (Bässler et al. 1974), so for the sake of simplicity only this stimulus amplitude will be discussed.

In a linear system one can see from the Bode plot of the open loop system whether the closed loop configuration is stable or whether oscillations occur. If the gain is equal to or greater than one for the frequency

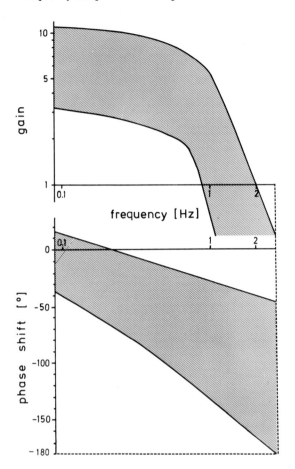

Fig. 2.17. Bode plot for a stimulus amplitude of 100 μm. Data from 10 animals. The data points lie within the *shaded area.* See text for definition of gain

with a phase shift of 180° (input and output are in antiphase), oscillations will show up in the closed system at this frequency. If, on the other hand, the gain is less than one at this frequency, the system is stable (Nyquist criterion).

As is readily apparent from the wave form of the tibia movement (see Fig. 2.16), this system is not linear. The Nyquist criterion nevertheless provides a reasonable first approximation. According to this criterion, the femur-tibia control loop is stable but has only a small phase reserve. A slight increase in phase shift (about 30°–40°) and/or gain should produce oscillations in the system. When a linear system oscillates, it does so with a theoretically unlimited amplitude. However, in a non-linear system stable oscillations of small amplitude may occur. Since this is a non-linear system, it is necessary to determine whether its phase reserve is indeed small. This can be done by experimentally increasing the phase shift and/ or gain to see if small amplitude oscillations arise in the closed loop system. If the above premises are valid, the frequency of these oscillations should lie between 1 Hz and 3 Hz.

If a system contains elastic structures such as muscles, its phase shift can be increased by coupling an inert mass to the system. Therefore the tibia from a restrained intact animal was attached to a weight suspended

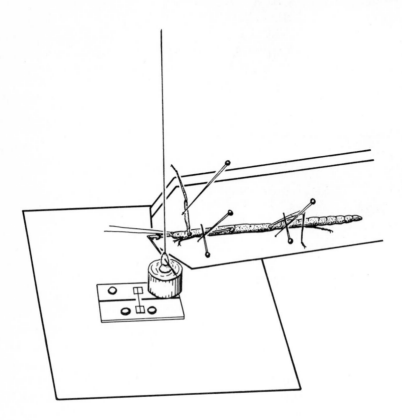

Fig. 2.18. Coupling of the tibia to an inert mass. The slot below the weight is used for recording the oscillations

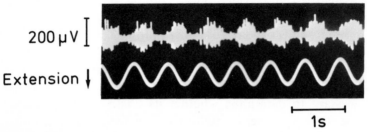

Fig. 2.19. Myogram from the flexor tibiae muscle during oscillations of a tibia that is coupled to an inert mass

from the ceiling (Figs. 2.18, 2.19). The resonance frequency of this pendulum was 0.5 Hz. The plane of movement of the tibia was horizontal. The tibia was fixed to the weight so that in its resting position no force was exerted on it, i.e., the weight functioned only as an inert mass. After disturbance of the animal, which increases the gain of the control system, long-lasting oscillations with frequencies between 1.2 and 2.5 Hz appear. These oscillations may be presumed to be feedback oscillations resulting from an increase in phase shift at high gain.

Results from three other experiments corroborate the hypothesis that the observed oscillations are in fact feedback oscillations: (1) No oscilla-

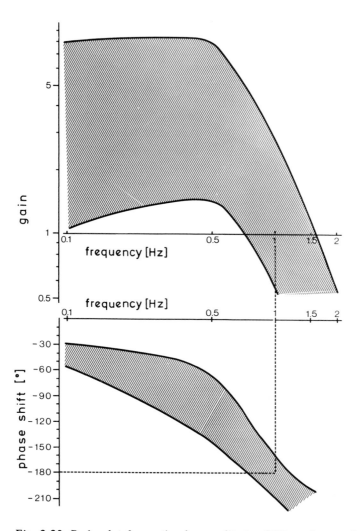

Fig. 2.20. Bode plot for a stimulus amplitude of 80 μm from a tibia that is coupled to an inert mass of 10 g. The data points from 10 animals lie within the *shaded area*

tions can be elicited by attaching an inert mass to a leg in which the receptor apodeme has been cut. (2) The electrical activity of the participating muscles can be recorded during such oscillations. Figure 2.19 is a sample myogram from the flexor tibiae muscle. It clearly demonstrates that the muscle is rhythmically active during each extension of the leg. (3) One can attach an inert mass to the tibia in the open loop system (defined stimulation of the chordotonal organ as in Sect. 2.4.2). The Bode plot for sinusoidal stimulation of the chordotonal organ shows that the inert mass indeed increases the phase shift. The amplitude-frequency characteristic is not appreciably changed (Fig. 2.20, unpublished). These results confirm the conclusion made from the Bode plot that the phase reserve of the system is very small (Bässler et al. 1974).

Thus, the femur-tibia control loop possesses another characteristic. Its phase reserve is so small that even a slight increase in phase shift produces long-lasting oscillations. Superimposed oscillations of 1–3 Hz should elicit autogenic oscillations, even without an inert mass, i.e., the control loop has a "resonance point." The existence of a resonance point can be demonstrated directly by attaching the tibia to an oscillator by means of a nylon thread which is held taut by a tiny weight hanging from its middle (simulating a very weak spring). The tibia moves to a notable extent only between oscillator frequencies of 1–3 Hz (Bässler et al. 1974).

2.5.4 Rocking After Ablation of Sense Organs – Definition of Central Oscillator and Program

The frequency of rocking corresponds to the resonance frequency of the femur-tibia control loop and rocking always appears when the gain of the control loop is high. To determine whether the control loops participate in the production of rocking, Pflüger (1977) recorded the electrical activity of the extensor and flexor tibiae muscles from one leg of an unrestrained animal during rocking. In the intact leg the extensor and flexor are alternatingly active during rocking (Fig. 2.15). This alternating rhythmic activity can still be recorded even after the control loop is opened by cutting the receptor apodeme of this leg and therefore feedback oscillations are blocked. The phase relationship between the electrical activity and the movement remains about the same as in the intact leg. This clearly demonstrates that the control loop of this leg is not responsible for rhythmic motor output. The rocking movement of one leg can apparently be produced without the participation of "its own" sense organs.

If the receptor apodemes of all six legs are cut and most of the other sense organs known to measure the positions of the other leg joints are ablated, it is much more difficult to elicit rocking. However, when rocking movements do occur, their frequency has clearly increased to an average of 3.6 Hz. The phase relationship between motor output and movement remains the same as in intact animals.

There is apparently a circuit in the CNS capable of producing a rhythmic motor output for all legs with nearly all the hallmarks of rocking and which in current usage can be termed a central oscillator. Since frequency changes occur when the periphery is intact, the rocking movements of all legs of an intact animal cannot be produced solely by this central circuit. Peripheral information must also comprise an important part of the system that produces rocking. The whole circuit that produces the rhythmic output and the way in which its individual components influence each other could itself be regarded as an oscillator, but I prefer to think of it as a program for rocking (Bässler 1967). The concept, program, is used here in the sense of a computer program (including hardware) and not for the sequence of motor output (see also the introduction to Sect. 4.2) as it is used by some authors. The "central oscillator" is most probably a part of the "rocking" program.

According to this terminology one has a "central oscillator" whenever the central nervous portion of a program for a rhythmic motor output is still capable of producing a rhythmic output after elimination of the periphery. The concept does not require that this part of the program also produce the fundamental rhythm under natural conditions, i.e., that

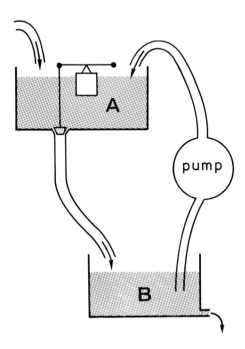

Fig. 2.21. A model to demonstrate the concept "central oscillator." A slow continuous stream of water flows into container *A*. When the water reaches a certain level, a float opens a valve which stays open until container *A* is empty. The water from *A* flows into container *B* where some of it leaks out and some of it is pumped back into *A*. The water level in *A* will oscillate rhythmically. If the pump is turned off (corresponding to the elimination of peripheral input), the system will continue to oscillate, but at a lower frequency

it governs the other parts of the program. It is more probably just a part of the program. The simple example in Fig. 2.21 should help to clarify this point.

Do the resonance properties of the femur-tibia loop participate in the control of rocking movements? To test this question the tibia insertion of the receptor apodeme was transplanted from the dorsal to the ventral side of the femur-tibia joint (crossing of the receptor apodeme, see Sect. 4.2.4.1; for surgical technique see Bässler 1967). After this operation the chordotonal organ signals the opposite of the actual movement of the tibia. Legs which have been operated on in this way can participate in rocking. Recordings of their electrical activity show that the phase relationship and duration of this activity is altered compared to that of the intact legs. The changes are such that one must assume that the control systems interfere with the output produced by the rocking program, which means that the resonance properties also contribute to the rocking movements.

According to this concept a program (or oscillator) of unknown location rhythmically modulates the appropriate motor neurons of all six legs. When the sense organs are intact, the frequency of this rhythmic modulation lies exactly at the "resonance point" of the femur-tibia control loop. Each control system can amplify the rhythmic excitation coming from the program when it is close to its resonance frequency. The control loops of the coxa-trochanter joint (see Sect. 3.3) may also have similar resonance properties (Pflüger 1976, 1977).

2.5.5 Further Support for a Central Oscillator
(see Sect. 2.5.4 for Definition)

The following findings support the hypothesis that the central nervous part of the rocking program can produce a rhythmic excitation all by itself: (1) When stick insects lie on their backs, their legs occasionally exhibit rhythmic movements especially at the femur-tibia joints. These movements have about the same frequency as rocking movements. All the legs on one side of the body are in phase (i.e., they all perform corresponding movements at the same time). The contralateral legs are exactly 180° out of phase (unpublished). (2) The ventral nerve cord of the thorax was totally denervated (for details see Sect. 4.2.6.1). Extracellular electrophysiological recordings were made from the stumps of nerves nl_2 (innervates the protractor coxae muscle) and nl_5 (innervates the retractor coxae muscle, for anatomical relationships see Fig. 6.6). The actual purpose of these experiments was to obtain information about the central nervous contribution to the control of walking. Since for anatomical reasons it is not yet possible to conduct such experiments on the actual "rocking" muscle pairs, the levator and depressor trochanteris muscles and the extensor and flexor tibiae muscles (see Sect. 6.1.2), these experiments were also evaluated in the context of rocking. After weak mechanical stimulation of the abdomen, rhythmic bursts sometimes occur which alternate exactly in the two nerves. Only the neurons that have the smaller impulses in the

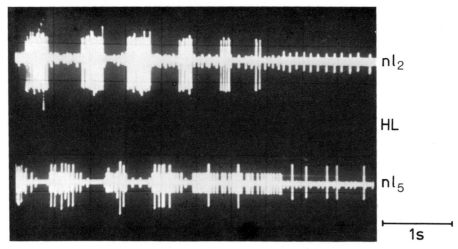

Fig. 2.22. Simultaneous recordings from the right nl_2 *(upper trace)* and nl_5 *(lower trace)* of the metathoracic ganglion in a denervated thoracic ventral nerve cord preparation

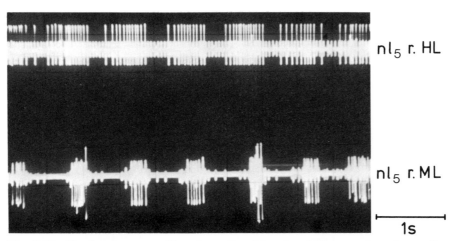

Fig. 2.23. Simultaneous recordings from the right nl_5 of the metathoracic ganglion *(upper trace)* and the right nl_5 of the mesothoracic ganglion *(lower trace)* in a denervated thorax preparation

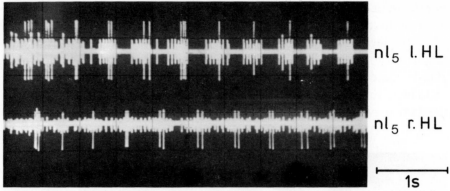

Fig. 2.24. Simultaneous recordings from a left *(upper trace)* and a right *(lower trace)* nl_5 of the metathoracic ganglion in a denervated thorax preparation

extracellular recordings are active (Fig. 2.22). This rhythmically alternating activity is very regular and can persist for 20 s. The frequency of this rhythm lies between 1.0 and 5.3 Hz, i.e., in the frequency range of rocking movements. The common inhibitor neuron also runs through these nerves and is usually inactive as during rocking movements (Bässler and Wegner, in prep.).

Recordings from the nl_5's of neighboring legs on a denervated thorax preparation show that the rhythmic activity of ipsilateral legs is always in phase (Fig. 2.23) whereas that of contralateral legs is in antiphase (Fig. 2.24).

The subcoxal joint, which is moved by the two muscles which were investigated, scarcely moves during rocking in intact animals. However, the experiments on denervated animals show that the motor neurons controlling this joint are rhythmically excited. Apparently the excitation is not amplified by the control system of this joint as it is in the femur-tibia joint, and therefore produces no appreciable movement of the subcoxal joint.

Rhythmically alternating activity in nl_2 and nl_5 of this kind has never been observed in decerebrated stick insects with denervated thoracic ventral nerve cords. Decerebrated animals that are otherwise intact do not rock. This supports the conclusion that the rhythmic activity of the denervated animal corresponds to rocking. It also makes evident the important role played by the supraesophageal ganglion in the production of rocking movements. Either this ganglion contains all or a part of the central nervous component of the rocking program, or the portion of the program located in the rest of the ventral nerve cord requires an input from the supraesophageal ganglion.

The proof for a rocking program and a resonance point of the control loop partly concludes stage 2 of the strategy for the investigation of rocking. The question of how the sense organs are integrated into the rocking program remains to be answered.

2.6 The Femur-Tibia Control Loop

The femur-tibia control loop is primarily movement sensitive. It possesses two characteristics, catalepsy (see Sect. 2.4.2) and a resonance point (see Sect. 2.5.3). The characteristic "catalepsy" is an independent behavior, whereas the characteristic "resonance point" plays a role in the generation of rocking. In the following chapters this control loop and the behaviors dependent on its characteristics will be analyzed in greater detail

2.6.1 Force Measurements on the Extensor and Flexor Muscles of the Tibia

The initial analyses of the femur-tibia control loop were based on force measurements from the tibia (Bässler 1974). Later studies by Storrer and Cruse (1977) in which the forces produced by the extensor and flexor muscles of the tibia were measured separately yielded similar results. For this reason only the latter experiments are discussed here.

Storrer (1976) and Storrer and Cruse (1977) used the following procedure for force measurements on the extensor and flexor tibiae muscles (see Fig. 2.25). The force generated by the flexor was measured directly from the cut flexor tendon. The tendon of the extensor was left intact. Since no muscles other than the flexor and extensor tibiae insert at this joint and the flexor tendon was cut, the force generated by the extensor muscle could be measured indirectly by a force transducer attached to the tibia. The forces produced by the extensor could be calculated from the length of the levers. Since the measuring accuracy of 1 mN for both muscles was not sufficiently sensitive to record the static portion, only the dynamic portion of the control system (see Sect. 2.4.2) could be investigated using this method. The stimulus was applied by moving the receptor apodeme as described earlier. The most important findings from these experiments are summarized as follows.

Figure 2.26 shows the results of *sinusoidal stimulation* at two different stimulus frequencies. The force amplitude and the phase shift can be determined from such curves as described in Sect. 2.5.3 (Fig. 2.27). It is

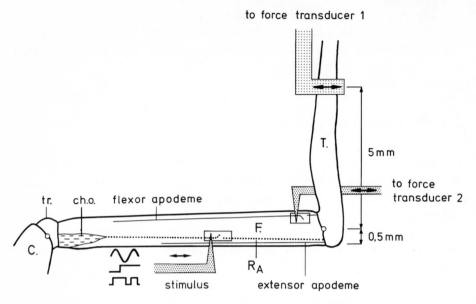

Fig. 2.25. The experimental set-up for measuring the forces generated by the extensor and flexor tibiae muscles in response to various stimulations (symbolized by different wave forms) of the chordotonal organ. *C* coxa; *ch.o.* chordotonal organ; *F* femur; *RA* receptor apodeme *(dotted line); T* tibia; *tr* trochanter. (Storrer and Cruse 1977)

obvious that the time course of force development is not sinusoidal for either muscle. The maximum force of the flexor was in all animals on the average about three times as great as the maximum force of the extensor.

At higher stimulus frequencies the force does not return to its baseline value, defined as the force generated by the muscle before the onset of stimulation. This discrepancy between force minimum and baseline increases with increasing stimulus frequency. This means that in the closed loop system moving the tibia back and forth with increasing frequency would prestretch the muscles more and more causing the muscles to become more rigid. Since in the closed loop system one of the two components of the force opposing a passive tibia movement is produced by the elasticity of the muscles (see Sect. 2.4.2), the opposing force increases as the muscle becomes more rigid. This explains why a passive sinusoidal tibial movement of higher frequency generates substantially more force than would be expected just from the control loop. An amplitude-frequency response curve that is arrived at using passive tibial movement first starts to decline at much higher frequencies than one that is determined on the open loop system where only the properties of the control loop itself are included. In the closed loop system the upper corner frequency is apparently just over 10 Hz (cf. Fig. 2.27). The phase shift is also

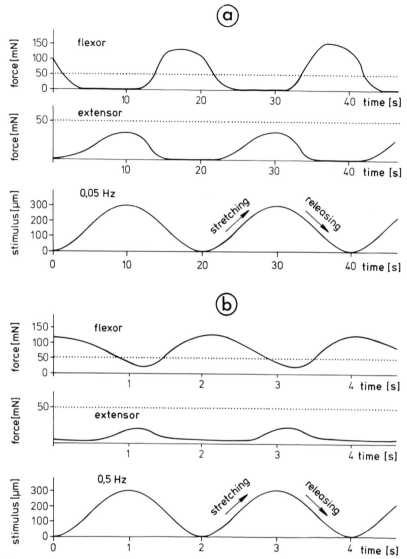

Fig. 2.26a, b. The time course of force development in the flexor and extensor tibiae muscles during two different frequencies of sinusoidal stimulation (*bottom trace* of a and b). The muscle force generated before onset of stimulation was used as the zero point for force. *Dotted lines* show the 50 mN force levels. (Storrer and Cruse 1977)

smaller than for the open loop configuration because the elastic forces have only a very small phase shift (Kemmerling and Varju 1981).

The muscle responses to "stretch-release" *step stimuli* are shown in Fig. 2.28. Shortening the interval between the release and the stretch has no effect on the maximum force generated by the extensor muscle. In

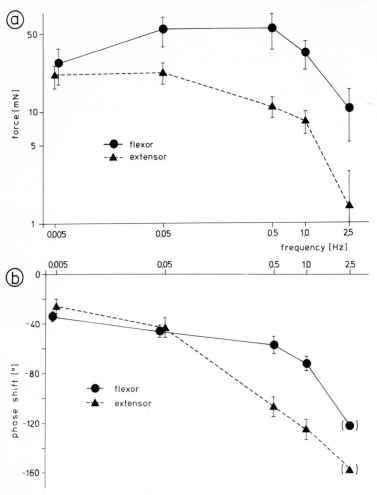

Fig. 2.27a, b. Force amplitude (**a**) and phase shift (**b**) as a function of stimulus frequency in the flexor and extensor systems. (Storrer and Cruse 1977)

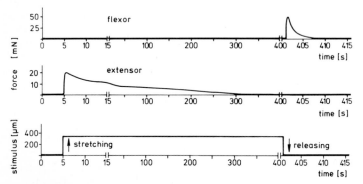

Fig. 2.28. The force generated by the flexor and extensor tibiae muscles in response to stretch-release step stimulation of the chordotonal organ as a function of time. Note that the middle of the time scale is compressed as marked by breaks in the abscissa. (Storrer and Cruse 1977)

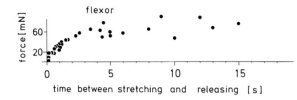

Fig. 2.29. The maximum force generated by the flexor tibiae muscle in response to a release step stimulus as a function of time following a preceding stretch step stimulus. (Storrer and Cruse 1977)

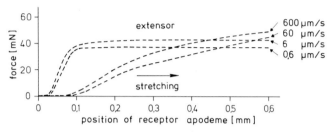

Fig. 2.30. Force curves from the extensor tibiae muscle for ramp stimuli (amplitude = 0.6 mm) with different slopes as a function of the position of the receptor apodeme. (Storrer and Cruse 1977)

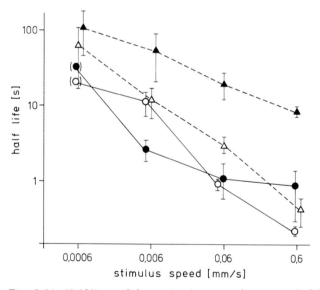

Fig. 2.31. Half-lives of force development *(open symbols)* and decline *(closed symbols)* for the flexor *(solid line)* and the extensor *(dashed line)* system as a function of stimulus speed. *Vertical bars* give the standard deviations. (Storrer and Cruse 1977)

contrast the maximum flexor response to a release step stimulus decreases if it directly follows (interval < 5 s) a stretch step stimulus which activates the extensor (Fig. 2.29).

Ramp stimuli (ramp and hold functions) were applied by stretching or releasing the chordotonal organ with constant velocity a set distance over varying periods of time. Force curves for stimuli of different slopes are shown in Fig. 2.30. Such force curves for both the extensor and the flexor reach a plateau during the stimulus ramp only if the stimulus velocity is low. After the set distance had been reached, the receptor apodeme was maintained in this position for 15 min, during which the muscle forces returned to baseline. Since force development during the ramp stimulus and force decline after the end of receptor apodeme movement approximate exponential functions, their half-lives can be estimated. Figure 2.31 shows that as the slope of the ramp increases both half-lives decrease, i.e., the tibia movement becomes faster. The force generated by the flexor tibiae muscle increases with ramp velocity to a maximum at a receptor apodeme movement of 0.5 mm/s, then declines with further increase of ramp velocity. The force generated by the extensor tibiae muscle is shown in Fig. 2.43.

2.6.2 Simulation of the Control System Based on the Results of Muscle Force Measurements (Cruse and Storrer 1977)

The muscle force data from Section 2.6.1 provided the basis for a simulation of the control system which is described briefly in the following. Since these measurements only encompassed the dynamic portion of the control system, the simulation only describes this part. The significance of such simulations and other methods of systems theory for behavioral physiology are discussed in Section 2.9.

First the whole system was conceptually partitioned into the "flexor system" and the "extensor system." Each of these two subsystems includes not only the properties of the corresponding muscle but also the properties of the chordotonal organ and the intermediate neural structures.

In each subsystem the input-output relationships for a step stimulus (Fig. 2.28) can be approximated by an exponential function of the form

$$f(t) = f_0 \left(e^{-\frac{t}{\tau_2}} - e^{-\frac{t}{\tau_1}} \right)$$

in which t = time, τ_1 = time constant of the low-pass filter (rise time constant), τ_2 = time constant of the high-pass filter (time constant of decline

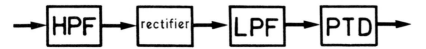

Fig. 2.32. Preliminary model for one subsystem of the feedback control loop. *HPF* high-pass filter; *LPF* low-pass filter; *PTD* time delay

or fall), and f_0 = a constant. Thus, each of the two subsystems can be treated as a band-pass filter with time constant τ_1 for the steeply rising branch (low-pass filter characteristic) and τ_2 for the falling branch (high-pass filter characteristic). A band-pass filter consists of a high- and a low-pass filter in series. Since each subsystem only responds to stimuli in one direction, a rectifier must also be included. The reaction time, which is about 50 ms for step stimuli, is realized by a time delay at the end of each channel. This preliminary model for each of the two subsystems is illustrated in Fig. 2.32 (for justification of the sequential arrangement see Cruse and Storrer 1977).

The values calculated from the step function half-lives (HL) according to

$$\tau = \frac{HL}{\ln 2}$$

are τ_1 = 0.22 s and τ_2 = 22 s for the extensor system and τ_1 = 0.3 s and τ_2 = 1.5 s for the flexor system. How well this preliminary model accounts for the results will be considered separately for the flexor and for the extensor system (for more detail see Cruse and Storrer 1977).

Flexor System: On the temporary assumption that the model is linear, the upper and lower corner frequencies ν_1 and ν_2 can be calculated from τ_1 and τ_2 using the equation

$$\nu = \frac{1}{2 \pi \tau}$$

This gives ν_1 = 0.6 Hz, which agrees well with the value estimated from the amplitude-frequency response (Fig. 2.27). Thus, a linear low-pass filter with τ_1 = 0.3 s describes the system within the measurement accuracy.

The lower corner frequency calculated from τ_2 is ν_2 = 0.1 Hz. According to the amplitude-frequency response the lower corner frequency must be less than 0.01 Hz. So the linear high-pass filter of the preliminary model fails to simultaneously describe both the step response and the amplitude-frequency response. Similar discrepancies are also found for ramp functions. With a linear high-pass filter the output amplitude should be proportional to the input slope; the half-life for response development and decline should be independent of the input slope; and both should be equal to the half-life of decline of the step response. None of these considerations is valid for the flexor system. Since the linear high-pass filter of

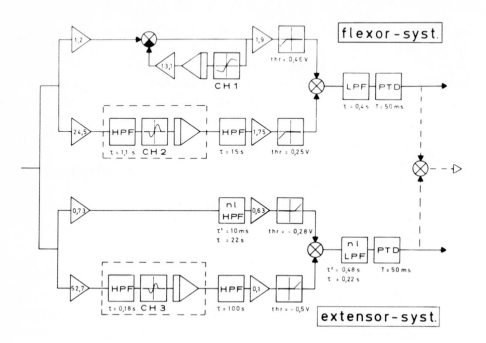

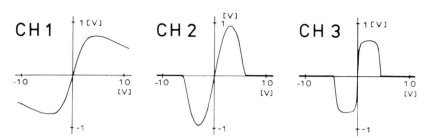

Fig. 2.33. A schematic diagram of the electronic model. *Triangles* represent amplifiers. Their gain is noted in the triangle. Integrators are shown by a *triangle with a rectangular base. Quartered circles* indicate summation of the input. A *filled quarter* denotes inversion of the corresponding input. *Thr* threshold; *PTD* pure time delay; *LPF* low-pass filter; *HPF* high-pass filter; *nl LPF, nl HPF* low- or high-pass filter with non-linear properties. (Cruse and Storrer 1977)

the original model did not describe the behavior of the actual system, it was replaced by a non-linear high-pass filter (see Sect. 2.7.1 for the significance of the properties of the non-linear high-pass filter for catalepsy).

There are two possible ways to simulate the non-linear properties of the high-pass filter. One possibility is to use at least two different high-pass filters with different properties. A switch (slope-window) preceeding these filters decides which one is available according to the slope of the stimulus (multi-channel model). The other solution uses only one channel constructed so that its time constant is a function of the slope of the input.

Cruse and Storrer (1977) chose the first alternative. One of the high-pass filters has a large time constant (τ = 15 s), i.e., the lower corner frequency is small. In front of this filter is a "slope-window" which only lets through stimuli with shallow slopes. The second channel has the properties of a high-pass filter whose time constant depends on the input amplitude but is always less than that of the other filter.

The performance of this model, shown in Fig. 2.33, is in good agreement with that of the actual system. The only features it does not account for are: the decrease in output amplitude with very steep ramp stimuli, the dependence of the output amplitude on the prestretching of the chordotonal organ, and the ability of the real system to change its gain over a wide range.

Extensor System. In the extensor system the same arguments hold for the partitioning of the high-pass filter of the original model into two channels. Channel 2 (see Fig. 2.33) is qualitatively the same as in the flexor system. However, a simple high-pass filter, which responds almost exclusively to stretch stimuli, can be used in channel 1 since in the extensor system the decline half-life after a step stimulus is independent of input amplitude. The use of a simple high-pass filter poses a problem because it predicts that the response to a stretch step stimulus is decreased if directly preceded by a release step stimulus. This behavior has only been observed in the flexor system where it can be explained in the following manner. A stretch stimulus cannot pass the rectifier to produce an output, but it does operate on the high-pass filter which precedes the rectifier. This means that it takes a certain amount of time for the output of the high-pass filter to return to zero after a stretch stimulus. If during this period it is followed by a release stimulus, the two responses are summed. Since the responses have opposite signs and the rectifier only passes positive values, the output amplitude is now smaller (see Fig. 2.34). In this context it is worth pointing out that this model demonstrates that the finding presented in Fig. 2.29 can be explained without invoking reciprocal inhibition between the

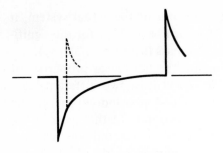

Fig. 2.34. See text for explanation

two subsystems. To simulate the behavior of the extensor the model uses a high-pass filter for channel 1 which responds almost exclusively to stretch stimuli.

The model presented in Fig. 2.33 satisfactorily describes the experimental results with the exception of the properties already mentioned for the flexor system.

2.6.3 Gain Control

The gain of the control loop can vary considerably for the same animal (Fig. 2.12). It is probably under central control since it is correlated with the animal's behavioral state. The simplest mechanism for such gain control would be via a direct input onto the motor neurons. However, if the motor neuron is spontaneously active, as is the SETi neuron, this input must also modulate the spontaneous firing rate. If this is how the system actually works, then the spontaneous SETi frequency should be correlated with its frequency after a standard stimulus irrespective of the behavioral state of the insect. To test this possibility SETi activity was registered over a period of several hours during which the chordotonal organ was presented with brief stretch-release step stimuli at irregular time intervals (5 min to 1 h). The animal was also disturbed at varying intervals (1–3 h). The prestimulus SETi frequency for the experimental animals varied between 0.5 and 20 Hz. No correlation between impulse frequency before and after stimulation was observed (unpublished). Schmitz (1980) repeated these experiments using a different experimental set-up in which animals stood with one hindleg on a small platform which could be moved to provide the stimulus (similar to set-up in Fig. 4.24). In these experiments there was much greater variation in the prestimulus SETi frequencies. Some correlation was found between the impulse frequency before and after stimulation, but it was only weak for any particular animal. Thus, the gain regulation of the control loop (at least for SETi) does not appear to be mainly due to an additional input onto the motor neuron. Force data from the flexor and extensor tibiae muscles (Bässler 1972c) suggest that this conclusion is valid for all the motor neurons (see also Sect. 2.3).

2.6.4 Influence of Other Sense Organs on the Input–Output Relationships of the Control Loop

In addition to the femoral chordotonal organ several *multipolar sense cells* near the femur-tibia joint monitor the position of this joint (see Sect. 6.2.3). Mechanical stimulation of each of these cells does not alter the spontaneous frequency of the SETi neuron (see Sect. 6.3.3; Bässler 1977a). They appear to have no influence on the control of the femur-tibia joint.

Stimulation of the *tension receptor* in the flexor tibiae muscle (see Sect. 6.2.3) slightly raises the SETi frequency. The tension necessary to elicit this effect has up to now only been attained in force measurements and not in experiments with an unrestrained tibia. Evidently, for an unrestrained tibia the femoral chordotonal organ provides the only sensory input to the femur-tibia control system. All conclusions on the behavior of the closed loop system are based on data from the unrestrained tibia. It is possible, however, that the force measurements were influenced to a certain extent by the tension receptor.

2.7 Evolution of Catalepsy and Rocking

2.7.1 The Femur-Tibia Control Loop in *Schistocerca*

In *Carausius* the characteristics of the femur-tibia control loop obviously subserve twig mimesis by contributing to the generation of catalepsy and rocking. It is possible that this system is so specialized in the stick insect that it has very little in common with analogous systems in "normal" insects. For this reason investigations of the femur-tibia control loop in another insect species with similar anatomical relationships in the femur are the logical next step in the study of this system. If the species is closely related to the ancestors of the phasmids, comparative studies have the added bonus of yielding information on the evolution of the system within the phasmids.

The migratory locust *Schistocerca gregaria* was selected for the comparative studies. The hindleg of this insect is highly specialized for jumping and can hardly be regarded as a forerunner of the phasmid leg. For this reason, after pilot studies by Rahmer (1972), investigations concentrated on the middle leg, which is an unspecialized walking leg in *Schistocerca*

(Ebner and Bässler 1978). The control system of the femur-tibia joint of this leg has a structure similar to that of *Carausius* and has been studied electrophysiologically (Burns 1974). Phylogenetic studies by Sharov (1971) suggest that the phasmids are derived from Orthoptera of the late Paleozoic era. These early Orthoptera had hindlegs with thickened femurs which could well have been jumping legs. Thus, within the suborder Orthopteroidea the control system of the *Schistocerca* middle leg could be considered more primitive than that of *Carausius*.

The experiments on *Schistocerca* were carried out in the same way as for *Carausius* (measurement of unrestrained tibia movement see Sects. 2.4.2 and 2.5.3). In the *open loop system* there is an amazing conformity with *Carausius*. The phase frequency response for sinusoidal stimulation is similar to that of *Carausius*. The amplitude-frequency response rarely has a gain greater than 1 and is thus considerably less than in *Carausius*. The upper corner frequency is also lower. The *Schistocerca* system has a greater phase reserve and is therefore very stable (Fig. 2.36).

The low gain is also evident in the step response. In addition the return movement toward the starting position is much slower. Calculation of the lower corner frequency from the decline half-lives gives values similar to those from the amplitude-frequency response. Each of the two subsystems can therefore be described by a model as shown in Fig. 2.32.

As expected, the *closed loop system* has properties which are only vaguely reminiscent of catalepsy (large amplitude of the fast phase, relatively high velocity in the slow phase, see Sect. 2.4.1). In *Schistocerca* even attachment of a large inert mass does not produce any long-lasting oscillations.

Assuming that the femur-tibia control system of all the legs of the ancestral phasmids in the late paleozoic had characteristics similar to those of the *Schistocerca* middle legs, one can infer that the control system of *Carausius* evolved primarily through a substantial increase in its gain. At the same time the complicated high-pass filter properties seem to have developed or at least intensified. These properties ensure that the system has a small time constant after a rapid flexion. The resulting rapid decline in the response is responsible for the small amplitude of the fast phase of catalepsy. The time constant is large for slow movements, i.e., the gain does not markedly drop during slow extension. In conjunction with the high gain this produces very low velocities during the slow phase of catalepsy. Thus, the complicated high-pass filter characteristics of the *Carausius* system seem to be an adaptation for the generation of catalepsy.

According to these findings the femur-tibia control system of *Schistocerca* middle legs possesses a preadaptation for the development of catalepsy and a "resonance point" eventhough this preadaptation is not expressed in the behavior of the closed loop system. The differences to *Carausius* are so slight that it is conceivable that the neuronal "wiring pattern" was maintained and only the synaptic properties had to be changed for the development of both these properties in *Carausius*.

2.7.2 Femur-Tibia Control Loop, Catalepsy and Rocking in *Extatosoma tiaratum* and *Cuniculina impigra*

As discussed in Section 2.7.1, the femur-tibia control system in the middle and hindlegs of *Carausius* may have developed from a system similar to the *Schistocerca* middle leg mainly by an increase in gain. The high gain is responsible for both peculiarities of the system, catalepsy and the small phase reserve. It could have developed under selection pressures favoring either catalepsy or a small phase reserve for the generation of rocking; or both selection pressures may have been present at the same time. A decision between these alternatives is made possible by investigating phasmids that show only one of these behaviors. *Extatosoma tiaratum* (see Fig. 2.35) shows rocking but very seldom catalepsy. *Cuniculina impigra* (syn. *Baculum impigrum,* see Fig. 2.35) exhibits a high frequency tremor but almost no actual rocking movements and has a very pronounced

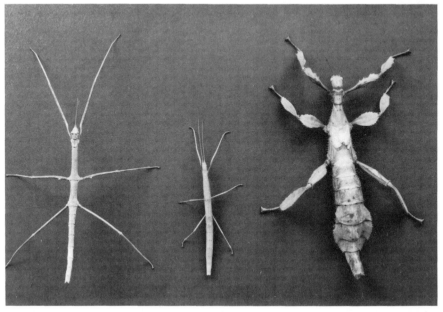

Fig. 2.35. Three phasmid species. *Cuniculina, Carausius,* and *Extatosoma* (left to right)

catalepsy. *Cuniculina* like *Carausius* belongs to the subfamily Lonchodinae, whereas *Extatosoma* belongs to the subfamily Podocanthinae. The discussion of *Extatosoma* is based on a study by Bässler and Pflüger (1979); that of *Cuniculina,* on one by Foth (1977). In both studies only females were used; in *Extatosoma,* the middle and hindlegs; in *Cuniculina,* only the middle legs.

In both species the femur-tibia control loop was first studied in the open-loop system (measurement of the position of the unrestrained tibia) using the same method as for *Carausius* (see Sect. 2.4.2). Figure 2.36 shows the Bode plots for both species in comparison with the ones for *Carausius*

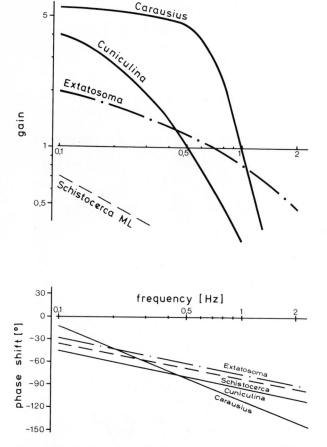

Fig. 2.36. Bode plots of the open loop femur-tibia control systems for different orthopterans, schematic presentation

and *Schistocerca.* The amplitude-frequency responses of *Extatosoma* and *Cuniculina* are similar to that of *Carausius.* Although the gain seems on the average to be somewhat smaller, comparison of the maximally record-ed amplitudes reveals that they have the same order of magnitude in all three species. The upper corner frequency is lower for *Extatosoma* and *Cuniculina* than for *Carausius.* The phase frequency response is about the same for *Cuniculina* and *Carausius; Extatosoma* has a smaller phase shift. The control systems for both comparative species have relatively large phase reserves (*Cuniculina,* due to its lower corner frequency; *Extatosoma,* due to its lower corner frequency and smaller phase shift). A "resonance point" of the kind found in *Carausius* is not present.

The step response can be used to determine the rise and decline half-lives, which can in turn be used to calculate the upper corner frequency. In *Cuniculina* the value for the corner frequency obtained in this manner agrees with that determined from the amplitude-frequency characteristic. (This value was not determined for *Extatosoma.*) The lower corner fre-quencies as calculated from the decline half-lives are in both cases as in *Carausius* (see Sect. 2.6.2) distinctly higher than those calculated from the amplitude-frequency response. Thus, complicated high-pass filter charac-teristics are also found in these two species.

The control loops of both species appear to possess the necessary pre-requisites for the appearance of catalepsy: a high gain and variable high-pass filter time constants which are large for stimuli with slowly rising slopes and small for stimuli with steeply rising slopes (see Sect. 2.7.1). *Cuniculina* does in fact show a distinct catalepsy. In accordance with the, on the average, smaller gain the return velocities of the slow phase in *Cuniculina* seem to be somewhat higher than in *Carausius.* When a leg has no tarsal contact, *Extatosoma* most often holds its femur-tibia joint in a flexed position (in contrast to *Carausius* and *Cuniculina* where the joint is normally extended in the absence of tarsal contact). For this reason the *Extatosoma* joint usually can be only extended passively during an experi-ment and, thus, does not show a very pronounced catalepsy. *Carausius* also exhibits only weak catalepsy after a passive extension (see Sect. 2.4.2). In the few cases where it was possible to passively bend the *Extato-soma* joint by an appreciable amount, catalepsy was clearly evident. Ap-parently *Extatosoma* has the ability to execute catalepsy but only rarely makes use of it.

The control of rocking in *Extatosoma* was investigated using the same methods as for *Carausius* (see Sect. 2.5.4). Cutting the receptor apodemes of one or all of the femoral chordotonal organs has as little effect on the phase relationships between the electrical activity of the muscles and the movement as did crossing of the receptor apodemes. This means that in

Extatosoma the control loops do not participate in the control of rocking. This is not surprising since these femur-tibia control loops do not have a "resonance point." In *Extatosoma* "rocking" can also be observed in an animal that is lying on its back: the tibias of all the legs move rhythmically, ipsilateral ones in phase with each other and in anti-phase with the contralateral ones. This phenomenon occurs much more frequently in *Extatosoma* than in *Carausius*. However, here too the sense organs can slightly influence the rocking frequency. After four receptor apodemes have been crossed and the remaining two cut, rocking frequency decreases. Thus, even in *Extatosoma,* rocking is not produced by an exclusively central program.

2.7.3 Hypotheses on the Evolution of Catalepsy and Rocking

The generation of rocking has apparently not exerted a selection pressure towards raising the gain of the control loop. Rocking in *Extatosoma* is produced without support from the control loops and only rarely occurs in *Cuniculina.* Both have a high control loop gain which, however, does not lead to a decreased phase reserve expressed as a "resonance point." This is due to a lowering of the upper corner frequency in both species and to a smaller phase shift in *Extatosoma.*

Selection pressures emphasizing catalepsy appear to be solely responsible for the change in control loop characteristics. In the course of evolution not only did the gain increase but the high-pass filter characteristics also changed in such a way as to favor the generation of catalepsy. Seen from this point of view, the small phase reserve is merely a by-product resulting from the high gain.

Other selection pressures apparently independently produced a program for rocking. There are only two conceivable ways to reconcile the simultaneous occurrence of a rocking program with a control loop having a high gain.

1. The frequency generated by the rocking program can be tuned to the "resonance point" of the control loops. This is an economical solution because the output amplitude of the program can be kept low and amplified later by the control loops. Rocking frequencies outside of the "resonance points" of the control loops would lead to complicated wave forms since each control loop would filter out that part of the output frequency which corresponds to its "resonance point."

2. The resonance characteristic of the control loop can be suppressed, either by making the control loop gain very small during rocking or by decreasing the upper corner frequency and/or the phase shift.

Case (1) seems to have been realized in *Carausius;* case (2) in *Extato-soma.*

2.8   Neural Basis of the Femur-Tibia Control Loop

2.8.1   The Motor Neurons of the Extensor Tibiae Muscle

The extensor tibiae muscle of *Carausius, Cuniculina* and *Extatosoma* is innervated by three units, a fast (FETi), a slow (SETi), and a common inhibitor (CI) neuron (for details see Sect. 6.3.3). Cobalt backfilling from nerve F2 was used to show the morphology of the neurons. The results are the same for all three phasmids. The position and form of FETi and SETi are shown in Fig. 2.37 and schematically in Fig. 2.38. In these preparations the soma of a third neuron can occasionally be seen near the ganglion midline and is probably the CI. Its location is indicated in Fig. 3.8.

We first used *Extatosoma* for the identification of these neurons be-cause in this species the somata can be seen with transmission microscopy (Fig. 2.39). The ganglionic sheath must be removed before insertion of a capillary electrode into the somata. The ganglia continue to function for some time after desheathing.

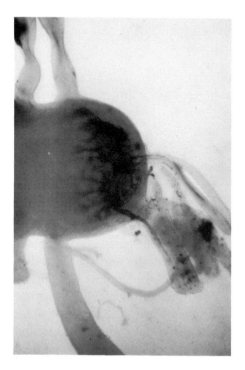

Fig. 2.37. The two excitatory motor neurons of the extensor tibiae muscle in the mesothoracic ganglion of *Carausius,* backfilled with cobalt from nerve F2 and intensified. The anterior end is up and the focal plane is between the cell bodies (ventral) and the main portion of the neuropil (dorsal)

anterior

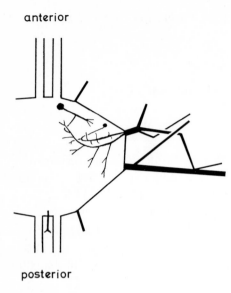

Fig. 2.38. Schematic dorsal view of the two excitatory motor neurons of the extensor tibiae muscle

posterior

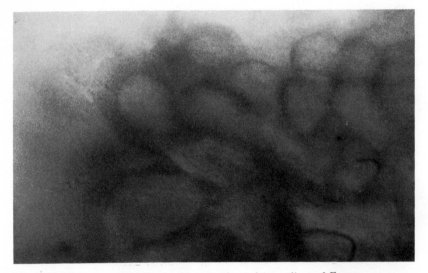

Fig. 2.39. Neuron cell bodies in the metathoracic ganglion of *Extatosoma*

Typical soma spikes are often observed when the electrode is inserted into a soma (the recorded spikes have small amplitudes and are quite slow due to the high resistance and capacity of the large amount of neuropil lying between the recording site and the spike-generating region). A sample recording is shown in Fig. 2.40. Electromyograms from the extensor tibiae muscle were made simultaneously with intracellular recordings to identify the neurons. The larger of the two neurons shown in Figs. 2.37 and 2.38

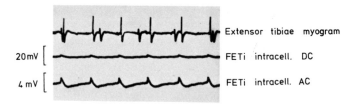

Fig. 2.40. Simultaneously recorded activity from the extensor tibiae muscle and the FETi motor neuron in the metathoracic ganglion of *Extatosoma*. The FETi was driven by injected current. *Small myogram spikes* are from the SETi; *large ones* from the FETi

proved to be the FETi, and the smaller, the SETi neuron (Bässler and Storrer 1980). The same results hold for *Cuniculina* (Bässler, in prep., see Sect. 2.8.3) and *Carausius* (Kittmann, unpubl.).

If current is injected into the FETi neuron causing it to fire continuously, the spontaneous frequency of the SETi is not affected. Apparently the SETi neuron does not receive information from the FETi (Bässler and Storrer 1980).

The position, shape, and absence of coupling between FETi and SETi neurons correspond to the equivalent neurons in *Schistocerca* pro- and mesothoracic ganglia but not to those in the metathoracic ganglion (Wilson 1979a, b). So far the findings are compatible with the hypothesis that the basic neuronal "wiring pattern" did not change during the evolution of the phasmid control loop.

In contrast to the conservative ganglionic parts of the motor neurons, their peripheral branchings are different in *Carausius* and *Schistocerca* (see Sect. 6.3.3). In *Schistocerca* the two motor neurons leave the ganglion via different nerves; in the phasmids, via the same nerve.

2.8.2 The Responses of FETi and SETi to Stimulation of the Femoral Chordotonal Organ

The simulation of the extensor part of the control system (Sect. 2.6.2) employs two separate channels, a "fast" and a "slow" one. The *Carausius* extensor tibiae muscle is innervated by a fast and a slow axon. Therefore, investigations were undertaken to test whether these neurons were the anatomical correlates for the output of the two simulation channels (Storrer, unpubl.).

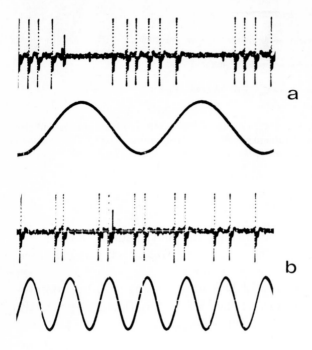

a

b

Fig. 2.41a, b. Recordings from the nerve F2 of SETi responses to sinusoidal stimulation (amplitude = 300 μm) of the femoral chordotonal organ at stimulus frequencies of 10 Hz (**a**) and 30 Hz (**b**). A single middle-sized spike from the CI can be seen in each recording. The small spikes that are barely visible above the noise level are sensory spikes. (Storrer, unpubl.)

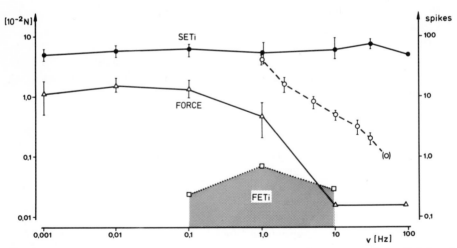

Fig. 2.42. Response as a function of stimulus frequency for extensor force (*left-hand ordinate* and *triangles*) and FETi and SETi responses *(right-hand ordinate)*. *Filled symbols* represent maximum spike frequency, *open symbols* spikes per cycle. *Shaded area* shows the frequency range where FETi spikes occur. (Storrer, unpubl.)

In these experiments the receptor apodeme of the femoral chordotonal organ was moved as described in Sections 2.4.2, 2.5.3, and 2.6.1. The force generated by the extensor tibiae muscle was measured as in Section 2.6.1. The electrical activity of nerve F2 (innervates the extensor tibiae muscle, see Sect. 6.1.3) was recorded with insulated steel wires, which were inserted through the cuticle so that they pressed the nerve against a trachea without mechanically influencing the muscle (Fig. 2.41).

SETi activity is synchronized with *sinusoidal stimulation* of the chordotonal organ over a wide range of stimulus frequencies. Its maximum spike frequency remains the same up to stimulus frequencies of 10 Hz (Fig. 2.42).

The behavior of the FETi neuron is more complex. Sinusoidal stimulation elicits activity only within a certain range of stimulus frequencies (Fig. 2.42). The number of FETi impulses is high at the beginning of sinusoidal stimulation and decreases rapidly as the number of stimulus cycles increases.

The FETi neuron also responds to *ramp stimuli* only within a narrow region of rise velocities. The frequency of SETi impulses increases with increasing rise velocity but decreases at very high velocities when the ramp becomes step-like (see Figs. 2.43 and 2.44). When the ramp slope is constant, activity in FETi and SETi increases with increasing ramp height and the degree of prestretching of the chordotonal organ. No absolute correlation between motor neuron activity and force development can be seen in the curves. In these experiments the inhibitory neuron is only rarely active. The occasional, usually single, spikes from this cell always appear directly following SETi activity (see Fig. 2.41).

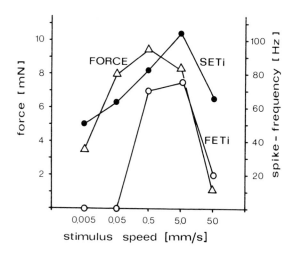

Fig. 2.43. SETi and FETi response and maximum extensor force as a function of the stimulus speed (rise velocity of ramp). The SETi values for stimulus speeds of 0.005 and 0.05 are the mean no. of spikes/s during the first 5 s of response; the other SETi values, during the ramp duration. The FETi values give the total no. of spikes. (Storrer, unpubl.)

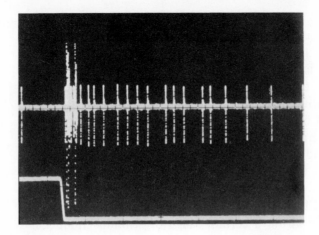

Fig. 2.44. Electrical activity in nerve F2 during fast ramp stimulation with an amplitude of 300 μm and a rise time of 20 ms. *Large spikes* are from the FETi; *smaller ones* from the SETi. (Bässler and Storrer 1980)

The FETi and SETi neurons cannot be stimulated separately with extracellularly applied current because they occupy the same nerve and their thresholds are practically identical. The following method was devised to study the effect of each neuron on the muscle. The muscle was bisected slightly distal to its midpoint. Its proximal portion is innervated almost exclusively by the FETi neuron; its distal portion by all three motor neurons (see Sect. 6.3.3). The force generated by each of the two muscle halves was recorded separately (Bässler and Storrer 1980). Figure 2.45 shows responses to step stimulation of the femoral chordotonal organ.

The amplitude and the half-lives of rise and of decline were measured from force development traces. Assuming that the forces produced by the FETi and SETi sum linearly, the characteristics of force development by the whole muscle can be roughly estimated. Accordingly during a step stimulus of 500 μm, the FETi neuron produces an average force of about 25 mN. The average half-life of rise is about 50 ms; of decline 100 ms. If the relationships were linear, these values would give an upper corner frequency of about 2 Hz and a lower corner frequency of about 1 Hz. This is in good agreement with Fig. 2.42. During the same stimulus the SETi neuron produces forces of about 10 mN. The upper corner frequency is about 0.1 Hz (significantly lower than that of the motor neuron); the lower corner frequency is about 0.0001 Hz.

These calculated values do not correspond with the values which were used for the two simulation channels. Thus, the FETi and SETi neurons cannot be regarded as anatomical correlates for the output of the two simulation channels (see Sect. 2.6.2). Impulse frequency measurements lead to the same conclusion.

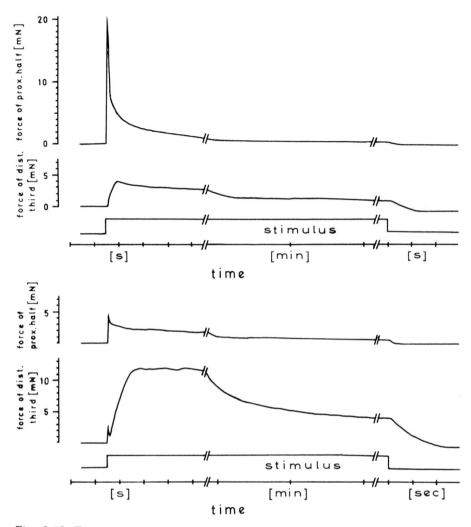

Fig. 2.45. Two extreme examples of force development by the bisected extensor tibiae muscle in response to step stimulation of the femoral chordotonal organ. See Fig. 2.44 for comparison of the motor neuron activity during a similar stimulus. FETi activity is strong in the *upper example* and weak in the *lower one*. (Bässler and Storrer 1980)

Continuous recording of the electrical activity from nerve F2 after a step stimulus reveal that the activity of the SETi neuron remains significantly raised even after more than one hour (Kittmann 1979). The incomplete adaptation of the system can thus be attributed to a neuronal property.

The extensor tibiae is a long muscle. A certain amount of time is necessary for electrical information to travel from the ganglion to the muscle.

The conduction velocities of the FETi and SETi axons were determined so that the contribution of conduction time to the reaction time of the control system could be estimated (Bässler and Storrer 1980). The conduction velocities are 3.3 m/s for the FETi axon and 2.4 m/s for the SETi axon. An FETi spike takes about 2 ms and an SETi spike about 5 ms to cover the distance from the ganglion to the middle of its region of innervation.

2.8.3 Alterations of the FETi and SETi Membrane Potential in *Cuniculina*

Extracellular recordings from its axon furnish information on the output of a neuron. However, its input—output relationship is necessary to determine the contribution of this particular neuron to the input—output relationship of the whole system. Intracellular recording provides the best means of capturing the sum of all inputs onto a neuron. We chose to use *Cuniculina* middle legs for these experiments because of the larger ganglia in this species. Soma recordings during stimulation of the chordotonal organ show that both neurons are depolarized by stretching and hypolarized by releasing of the chordotonal organ. After the end of both a hyper- and a depolarizing stimulus, the membrane potential returns almost to its original value, however, a small difference persists (Fig. 2.46). The rapid decline of hyperpolarization explains why the response to a "stretch" stimulus is the same whether or not it is immediately preceded by a "release" stimulus.

During rampwise stretch stimulation of the chordotonal organ (constant stimulus amplitude) the amplitude of the depolarization depends on the rising slope of the stimulus (the steeper the rise, the greater the de-

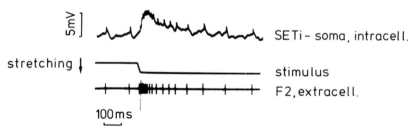

Fig. 2.46. Sample recording from the SETi soma during rampwise stretch (amplitude = 300 μm) of the femoral chordotonal organ in *Cuniculina*. *Lower trace* shows a simultaneous F2 recording

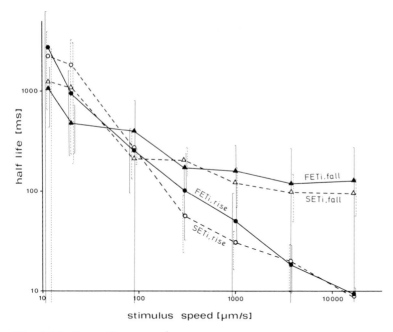

Fig. 2.47. The half-lives of rise and decline (fall) of depolarization of FETi and SETi somata as a function of stimulus speed in *Cuniculina*. Stimulus amplitude = 300 µm. Each *symbol* is the mean of 10 measurements. *Vertical bars* give the standard deviation

polarization). The half-lives of rise and fall decrease with increasing slope (Fig. 2.47). This non-linearity, which was also observed during the force measurements, is evident in both neurons, and their average values are almost the same for both neurons. Hyperpolarization in response to a release stimulation of the chordotonal organ also depends on the rising slope of the stimulus, and its rise and decline half-lives also decrease with increasing stimulus slope. Again the average values for both neurons are practically identical. Thus, for ramp stimuli the response characteristics of the FETi and SETi neurons are remarkably similar.

The steepest ramp stimuli closely approximate step stimuli, and the responses to them can be treated as step responses. The rise and decline half-lives can be used to calculate the time constants of the low- and high-pass filter, τ_1 and τ_2, which can in turn be used to calculate the corner frequencies (for details of calculations, see Sect. 2.6.2). For both neurons τ_1 is about 12 ms (equivalent to an upper corner frequency of about 14 Hz); τ_2 is about 150 ms (equivalent to a lower corner frequency of about 1 Hz). The upper corner frequency of both somata is, thus, clearly higher than that of the system as a whole. The lower corner frequency for the FETi neuron agrees approximately with values obtained from force mea-

surements and extracellular recordings; but is distinctly higher for the
SETi neuron.

For *sinusoidal stimulation* the response of the SETi neuron remains
relatively constant over several stimulus cycles. However, in the FETi
neuron the membrane potential change decreases in amplitude during
continued stimulation. Also, the average membrane potential shifts in the
direction of hyperpolarization (Fig. 2.48). These two processes are also
seen with repetitive ramp stimulation. Neither of these processes is very
extensive, but since both changes are in the same direction, they lead to a
rapid decrease in the number of FETi spikes per stimulus cycle.

The amplitude of the membrane potential change (only the first two
cycles for the FETi neuron) and the phase shift between stimulus and response were measured from records like those in Fig. 2.48. In Fig. 2.49
these values are given in the form of amplitude- and phase-frequency
characteristics. In the lower frequency range the phase-frequency response
exhibits the positive phase shift which is typical for a velocity-sensitive
system (high-pass filter). The large phase shifts shown by the system as a
whole are not evident here.

The amplitude-frequency response is similar for both neurons. The
restriction of FETi firing to a narrow frequency band is apparently mainly due to the FETi neuron's high threshold.

The lower corner frequency calculated from the step response has a
higher value than when it is derived from the amplitude-frequency response. This applies to both neurons. The high-pass filter properties in the
input of both neurons therefore are nonlinear in the same way as those of
the whole system. Both neurons seem to receive about the same input.
Thus, the frequency selectivity of the FETi neuron cannot be attributed
to its receiving input only from certain units of the chordotonal organ,
which possess the corresponding frequency selectivity.

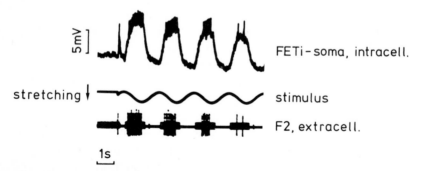

Fig. 2.48. Electrical activity of the FETi motor neuron and the nerve F2 during
sinusoidal stimulation of the femoral chordotonal organ

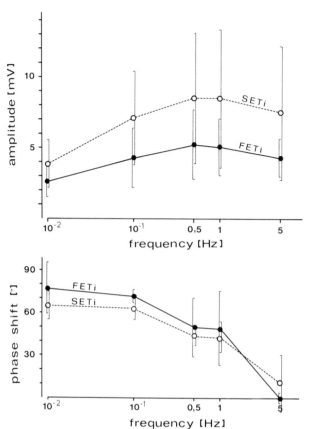

Fig. 2.49. Amplitude and phase shift of the membrane potential change in FETi and SETi somata as a function of the frequency of sinusoidal stimulation. *Data points* are the mean of 10 measurements; *vertical bars* give the standard deviation

2.8.4 Neuronal Equivalents of the Elements of Control-Loop Simulation

According to the results that are presently available, the various elements of the extensor part of the control system (shown in general form in Fig. 2.32) have the following neuronal equivalents (applying the results from *Cuniculina* to *Carausius*) (Fig. 2.50).

Reaction Time: The response to ramp stimuli is first evident in the soma of the motor neurons after about 10 ms. The SETi neuron, which is normally spontaneously active, responds immediately to this change with a change in its spike frequency. In the FETi neuron a certain time is necessary (1–2 ms) before the threshold for spike triggering is exceeded. The conductance time to the muscle takes a further 2 ms for the FETi neuron and 5 ms for the SETi neuron. Thus, in total no more than 20 ms can be attributed to neuronal processes, and the remaining 30 ms of the total

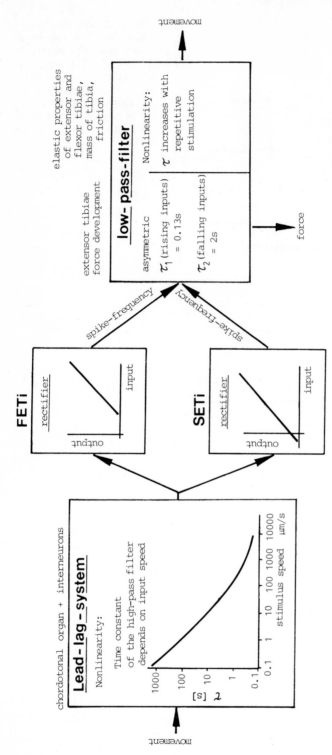

Fig. 2.50. The extensor part of the control system in *Cuniculina*. This model does not describe: habituation of the motor neurons, gain control, and time delay. *Lead-lag-system*: high pass filter with tonic component

reaction time (about 50 ms) must be due to processes in the muscle (end-plates, electromechanical coupling, force development).

Low-Pass Filter: The decisive low-pass filter for the forces elicited by the SETi neuron is represented by properties localized within the muscle. In the FETi neuron the low-pass filter is comprised not only of muscle properties. It is also to some extent a property of the neuron, resulting from the small decline in membrane potential amplitude at higher stimulus frequencies combined with the high threshold for spike generation. During sinusoidal stimulation the decrease in response intensity with repetitive stimulation also affects the low-pass filter characteristics (for further properties of the muscle see high-pass filter).

Rectifier: Since both neurons can respond with depolarization or hyperpolarization, the rectifier feature is restricted to their spike-generating mechanisms.

Non-Linear High-Pass Filter: The non-linear high-pass filter properties of the model are determined at several different neuronal levels. The dependence of the time constants on stimulus slope is already included in the motor neuron input and is, thus, a property of components that precede it in the circuit (interneurons and chordotonal organ). The time constants for the motor neuron input for higher input slope, however, are about one order of magnitude lower than for the muscle forces. Correspondingly, the lower corner frequency determined from the amplitude-frequency response is substantially higher for the motor neuron input than for the extension force. The "neuronal" high-pass filter has, therefore, smaller time constants for steep-slope inputs than the "model" high-pass filter. In reality, muscle properties seem to be responsible for slowing down the motor neuron output because the SETi spike frequency has about the same characteristics as the motor neuron input. An increase in spike frequency leads to a much higher force time constant than does a decrease in spike frequency. Thus, the real muscle low-pass filter is asymmetric. It has a small time constant for increasing and a big time constant for decreasing input (see also Sect. 2.8.2). This might be due to "catch" effects (i.e., fewer spikes per unit time may be necessary for the maintenance of a certain force than for its production). The "real" system for *Cuniculina* is given in Fig. 2.50.

For the FETi neuron additional factors contribute to the discrepancy between the lower corner frequencies as calculated from the step response and from the amplitude-frequency response. Since the FETi response decreases with repetitive stimulation, sinusoidal stimuli (responses were analyzed only after the first or second stimulus cycle in force and movement measurements) elicit less activity than step stimuli. Also, the higher spike threshold in the FETi neurons results in the lower corner frequency of spikes always being higher for the FETi than for the SETi neuron.

2.9 Significance of Systems Theory (Cybernetic) Methods for the Strategy Used Here

In the preceding analyses the methods of systems theory were employed in three ways:

1. They were used to demonstrate that a particular behavior is a characteristic of a quantitatively described system, i.e., that this system alone is capable of generating the behavior in question. The proof that catalepsy and "resonance point" are characteristics of the femur-tibia control system are examples of this kind of application. For such an analysis other methods can replace systems theory only at considerable inconvenience if at all.

2. Systems theory methods were indispensable for identifying certain system parameters, e.g., the realization that the high control loop gain is responsible for both peculiarities of the system (catalepsy and resonance point) and the discovery of the complicated high-pass filter properties in phasmids.

3. These methods could be used to construct a model to simulate a particular behavior, in our case, that of the femur-tibia control loop during various kinds of stimulation of the femoral chordotonal organ (see Sect. 2.6.2). If this model could be shown to be the only one possible, then the same formalism would apply to both the model and the actual system. The actual system should then be mathematically transformable into the model. But, it is not generally possible to prove that no other model can describe the behavior in question. Therefore such models have only a heuristic value. They can only stipulate how a system could function, not how it actually does. Despite this limitation such models are useful for suggesting new lines of experimentation. In our studies the simulation from Section 2.6.2 gave us further impetus to investigate the properties of the FETi and SETi neurons. Unlike (1) and (2), the construction of a heuristic model is not a necessary step in the investigation of a system; however, it is likely to simplify the formulation of the appropriate experimental questions.

2.10 Arousal

Arousal generally refers to the process of awakening, the transition from a "sleeping" state to an active one. Physiologists, however, also use this term to describe the shift of a population of neurons into a state of higher reactivity by a nonspecific stimulus. Often it is assumed that an externally visible "arousal" is accompanied by a simultaneous increase in reactivity. This assumption is not valid for *Carausius* because of the following considerations. (1) The gain of the femur-tibia feedback loop, which is a measure for reactivity, is not clearly correlated with the depth of thanatosis, a measure for arousal (see Sect. 2.6.3 and 2.3). (2) Disturbance as an arousing stimulus increases the gain of the femur-tibia feedback loop but only as long as the insect remains immobile. Once a restrained insect becomes active, the gain of the feedback loop decreases to the extent that it can no longer be measured (see Sect. 3.2.2). In a walking stick insect the gain of the feedback loop is low (see Sect. 4.2.7.1). (3) There are apparently two different kinds of arousing stimuli and states of arousal. On the one hand, severe disturbance leads to brief rapid movements, in some cases even to escape walking. After these movements end, the gain of the femur-tibia feedback loop is very high. Dimming of the ambient light level elicits, on the other hand, spontaneous movements followed by a low feedback loop gain. (4) The intensity of the response of the retractor unguis muscle (flexes the tarsus) to a defined tapping stimulus is decreased as much by "arousing" mechanical, chemical, or thermal stimuli as by CO_2 narcosis (see Sect. 3.1).

The appearance of common cause and correlation invoked by the double usage of the concept "arousal" can be circumvented in two alternative ways. One can eschew altogether the use of the term "arousal" and use separate concepts for the externally visible "awakening" and for increased reflex excitability, or one can treat thanatosis as itself an aroused state so that the transition from thanatosis to another active state would no longer be considered arousal. I prefer the first alternative since in our analysis thanatosis is treated as a variable instead of a state (see Sect. 2.3) and because there is a continuum not only between different feedback loop gains but also between degrees of thanatosis.

3 Other Behaviors of the Stationary Animal

3.1 Claw Flexing (Walther 1969)

Since the position of all the tarsal joints is controlled by one muscle, the retractor unguis (see Sect. 6.1.2), direct inferences on the state of the muscle can be made from movements of these joints.

The tarsus of a resting stick insect is usually held in a slightly curved position when the leg has no tarsal contact. The end segments of the tarsus often make irregular, small amplitude (maximally half the width of the tarsus) movements at a frequency of about $80-120$ min^{-1}.

In response to an unspecific stimulus the tarsus flexes rapidly then slowly relaxes to its starting position. This behavior may correspond to the insect's letting itself drop from its perch when disturbed (see Sect. 2.1).

In the experimental situation one hindleg is fixed so that it has no tarsal contact. The other legs rest on a treadwheel. A movement of the tarsus of the restrained leg can be released by a standardized tapping stimulus. When the animal is walking, the latency between stimulus and response increases and the starting position of the tarsus is less flexed (see Fig. 3.1), i.e., the muscle tone is less than when the animal is standing still. Cutting the connectives between the meso- and metathoracic ganglia

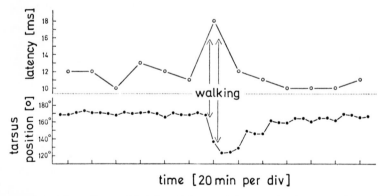

time [20 min per div]

Fig. 3.1. The effect of brief walking activity on the latency of tarsal response and muscle tone as gauged by the degree of tarsus bending in the dark. (Walther 1969)

abolishes these walking effects. An increase in latency with a simultaneous decrease in muscle tone can also be elicited by mechanical stimulation of the abdomen; mechanical, chemical, or thermal stimulation of the tarsus; or CO_2-narcosis. These changes may also occur spontaneously. The significance of these findings for the concept "arousal" is discussed in Section 2.10. Eventhough latency and muscle tone always change simultaneously, there is no significant correlation between their absolute values.

A periodic flexing of the hindleg tarsi (waving) is often shown by animals in which the connectives between meso- and metathoracic ganglia have been cut. This movement can be released by stimulation of the tarsus and CO_2-narcosis but not by abdominal stimulation. For further details on this movement see Walther (1969).

3.2 Active Movements of the Femur-Tibia Joint in Restrained Animals

3.2.1 Quantitative Description of the Movement

For the quantitative characterization of active movements the animals were prepared as shown in Fig. 2.7 (without cutting open the femur) and the experiments were performed under daylight conditions. Only the femur-tibia joint of one leg could move freely. Brief mechanical stimulation of the abdomen often elicited active movements, which were usually mainly in one direction (either flexion or extension). Sometimes back and forth movements occurred, but these rarely persisted for very long. The maximal velocity of these movements was $1800°$ s^{-1} for flexion and $1200°$ s^{-1} for extension, clearly faster than movements elicited by stretching or releasing the femoral chordotonal organ. The speed of active movements was on the average somewhat slower for legs with cut receptor apodemes (opened feedback loop) than for intact legs. Since a feedback loop always has a damping effect, the simplest explanation for these findings is that the feedback loop does not function during the actual active movement. There may even be a positive feedback. If the feedback system were functioning normally, the movements of legs with cut receptor apodemes would be faster than those of intact legs.

After an active movement the tibia usually returns slowly towards the position it was in before the active movement. The velocities of such return movements have the same order of magnitude as those of the slow phase of catalepsy. Indeed, when immediately before or after an active movement catalepsy is induced by passive flexion of the joint (see Sect. 2.4.1), there is a significant correlation between the return velocities after an active movement and in the slow phase of catalepsy. Legs with cut receptor apodemes usually return relatively quickly to the starting position after an active movement. All these findings lead to the conclusion that the same conditions prevail after both active and passive movements, i.e., the femur-tibia feedback loop is operating in both cases and the return movement after an active movement can be regarded as catalepsy (Bässler 1973).

3.2.2 Stimulation of the Chordotonal Organ in the Active Animal

The conclusions that have been discussed up to now are further corroborated by experiments in which the chordotonal organ was stimulated during active movement. The recording arrangement was the same as shown in Fig. 2.7. Experiments were first performed on decerebrate animals because they remain active longer (see Sect. 2.2). Either the tibia movement or the force exerted by the tibia on a force transducer was recorded. Figure 3.2 shows the response to fast rampwise back and forth movements of the receptor apodeme. During the actual active period (recognizable from the movements between two stimuli) each stretch of the chordotonal organ elicits a marked flexion, whereas release does not elicit a clear response (Fig. 3.2 a). After the end of the actual active period active flexion occurs only at the end of a stretch stimulus and the amplitude of the movement gradually decreases (Fig. 3.2 b). This is followed by a period (Fig. 3.2 c) during which stretch and release elicit flexion. Finally

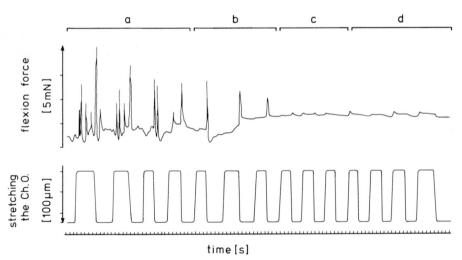

Fig. 3.2. Flexion force in response to rapid back and forth ramp stimulation (amplitude = 250 μm) of the chordotonal organ *(Ch.O.)* in the middle leg of a decerebrated animal. The transition from the active *(a)* to the inactive *(d)* state is unusually fast

the system responds as in the inactive animal with the gain increasing gradually (Fig. 3.2 d). Normally the transition from active to inactive state lasts considerably longer. Sometimes there is no reaction.

When intact animals are induced to perform long-lasting active movements by continuous abdominal stimulation, they show the same behavior with one essential difference. The transition from the active state (during which stretch elicits flexion; and release, no response) to the inactive state (where stretch elicits extension; and release, flexion) occurs quite rapidly (always in less than 1 s, often very abruptly). The gain of the feedback system is instantly and fully present. Figure 3.3 shows such transitions during sinusoidal stimulation (Bässler 1973, 1974).

Force measurements recorded separately from the extensor and flexor tibiae muscles (see Sect. 2.6.1) show that the flexion response to a stretch stimulus in the active state is brought about by deactivation of the extensor and activation of the flexor tibiae muscle (Storrer 1976).

Hence, in a restrained insect the femur-tibia feedback loop is turned off during active movements and is often replaced by a positive feedback restricted to flexing movements. In intact animals the feedback loop resumes operation at high gain immediately after the period of active movement. In decerebrate animals the transition to the inactive state takes much longer. This corresponds to the gradual increase in feedback loop gain following spontaneous active movements (see Fig. 2.12).

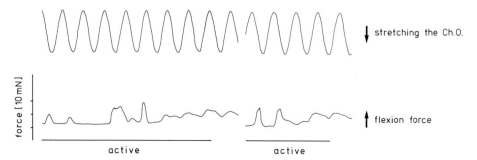

Fig. 3.3. Transitions between active and inactive states as measured by flexion force during sinusoidal stimulation of the chordotonal organ. Stimulus amplitude = 500 μm; frequency = 0.5 Hz. (In the active state the other legs are also moving)

3.2.3   Reflex Reversal, Program-Dependent Reaction

The response to a stretch stimulation changes in intensity (see Sect. 2.4.2) and direction (see Sect. 3.2.2) depending on the animal's state of activity. A reaction in which the direction and/or intensity depends on the motor program in use is termed a program-dependent reaction. Program-dependent reactions will be encountered several times in the following sections (see Sects. 4.2.4.2, 4.2.6.3, 4.2.7.1, 4.3.2, 4.3.5, 4.5).

Does such a "reversed" reflex use the same motor neurons as the "normal" reflex? In order to answer this question the activity of the SETi neuron was recorded extracellularly from nerve F2. Figure 3.4 shows that in the inactive insect a steep rampwise stretch of the chordotonal organ

Fig. 3.4. Activity of the slow extensor tibiae motor neuron during rampwise stimulation (shown in *top trace* with up denoting stretch) of the femoral chordotonal organ. *Vertical lines* represent spikes. *Second trace* inactive animal; *third, fourth* and *sixth traces* active decerebrated animals; *fifth trace* active intact animal. *Bottom trace* is the time scale in 100 ms units. (See also Fig. 4.20)

Fig. 3.5. Activity of the slow extensor tibiae motor neuron during triangular stimulation *(top trace)* of the femoral chordotonal organ in decerebrated animals. *Second trace* inactive animal; *third trace* active animal

raises SETi activity for some time while a corresponding release decreases SETi activity. In the active animal a steep rampwise stretch inhibits SETi activity for a short period during which the flexor motor neurons are very active. A release of the chordotonal organ elicits no response unless the animal happens to be in transition to the inactive state (Fig. 3.4, bottom recording). Stepwise stimulation of the chordotonal organ only rarely causes inhibition of the SETi neuron. Rampwise stimulation often reveals the most striking inhibition (see Fig. 3.5) (Bässler 1976).

The "active" response and the response triggered by the feedback loop have distinctly different time courses. This leads to the conclusion that reversal is not due to a simple change in sign. Rather, in both cases at least part of the information must flow over separate channels which have different properties. Each makes use of different units of the chordotonal organ and/or different interneurons but uses the same motor neurons.

3.3 Control of the Coxa-Trochanter Joint

The coxa-trochanter joint also possesses a feedback loop. Its receptor organ is the hair plate BF1 (see Sect. 6.2.2) as shown by the following observations: (1) If one bends the hairs of BF1 with a pin, the leg moves downward. (2) A standing stick insect can support four times its body weight without collapsing. After the BF1 hairs are removed from all legs, the animal is unable to support its own body weight (Wendler 1964).

This feedback loop has certain similarities to the femur-tibia feedback loop. (1) It appears to have a dynamic component as evidenced by the following observations. (a) Graphs of trochanter deflection as a function of applied force reveal a distinct hysteresis for increasing and decreasing forces (Fig. 3.6, Wendler 1964). (b) Plots of the firing frequency of the slow depressor trochanteris motor neuron against joint angle during slow

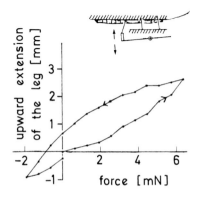

Fig. 3.6. The position of a right hindleg as a function of increasing and decreasing force. When leg extension = 0, the ventral body surface is 1 cm above the tarsus. The tarsus of the deflected leg is 2 cm behind the tarsus of the middle leg which is standing vertically to the body. The load was raised or lowered at intervals of two minutes. (Wendler 1964)

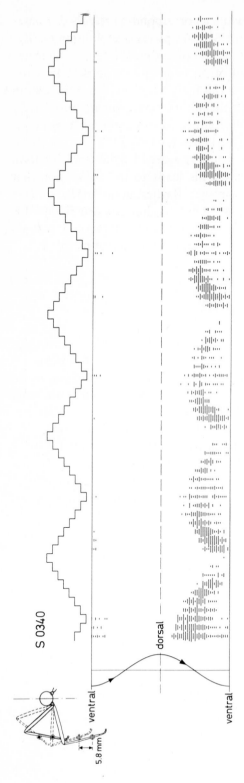

Fig. 3.7. The activity of a fast levator trochanteris motor neuron during sinusoidal movement of the coxa-trochanter joint at a frequency of 1 Hz. The difference between highest and lowest tibia positions was 5.8 mm (see *inset*). A step-like deflection of the joint was superimposed on the sine wave. *Upper trace* the average value of the sinusoidal stimulus. *Lower trace* spike distribution relative to the phase of the stimulus. (Wendler, unpubl.)

upward and downward trochanter movement resemble Fig. 2.11. Since this difference between upward and downward movement persists even at very low movement velocities (Wendler 1972), the dynamic component must be sensitive to very low velocities. (2) The system appears to have only a small phase reserve. Attachments of an inert mass to the femur (similar to the method used for the femur-tibia joints as described in Sect. 2.5.3) leads to long-lasting regular oscillations following disturbance (unpublished). This property may contribute to the production of rocking. (3) The activity of the slow depressor trochanteris motor neuron follows the stimulus up to a very high frequency range (over 10 Hz, Wendler 1972). (4) The fast and slow motor neurons seem to have a similar input but different thresholds. Wendler (unpublished) showed this for a fast levator trochanteris motor neuron: he moved the joint sinusoidally up and down at a frequency of 1 Hz, raising the fast motor neurons above its threshold. The average value of the sine wave was shifted up and down in steps (Fig. 3.7). The phase of the fast levator neuron appears to be shifted forward when the leg moves downward. Graphs of the number of spikes against the average position of the leg resemble Fig. 2.11.

3.4 Control of the Subcoxal Joint

3.4.1 Description of the Response

The receptors serving this feedback loop are the hair plate BF2 (Wendler 1964), an as yet unidentified receptor at the trochantin (see below) and, to a lesser extent, the hair rows on the coxa (Bässler 1965) (for anatomy see Sects. 6.1.1 and 6.2.1). When an increasing followed by a decreasing force is exerted on the joint, the graph of deflection against applied force reveals a hysteresis similar to that of Fig. 3.6 (Wendler 1964). This indicates the presence of a dynamic component as does the fact that the joint deflection caused by a particular load increases with time (Bässler 1965).

3.4.2 The Motor Neurons of the Retractor Coxae Muscle

The subcoxal joint is moved mainly by the retractor and protractor coxae muscles. The motor neurons of the retractor coxae muscle were used for the initial analyses because they are more easily identified in extracellular recordings. The adjacent portions of the nerve (nl_5) leading to the retractor contain only retractor motor neurons; whereas nerve nl_2, which innervates the protractor coxae, contains many other axons.

According to Graham and Wendler (1981a) subdivisions R_a and R_b of the retractor coxae muscle are innervated by at least five motor neurons: the common inhibitor (CI), a slow neuron (slow retractor coxae = SRCx), two intermediate or semifast neurons (semifast retractor coxae = SFRCx), and a fast neuron (fast retractor coxae = FRCx) (see Sect. 6.3.1). Igelmund (1980) describes a further semifast neuron and for R_c a semifast and a fast neuron which do not innervate R_a and R_b.

Cobalt backfills from branches of nerve nl_5 adjacent to the muscle reveal three to seven darkly stained neurons in the ganglion (probably the excitatory motor neurons) and a lightly stained soma. This must be the CI unit since cobalt backfills of the extensor nerve F2 (see Sect. 2.8.1) show a soma in the same position (see Fig. 3.8; Graham and Wendler 1981a; Igelmund 1980).

Identification of single neurons is not possible with soma recordings because (1) the somata lie very close together and (2) apparently no distinct spikes can be recorded intracellularly from the soma (Graham and Godden, in prep.). Recordings have been made from the neuropil of neurons SRCx, SFRCx1 and SFRCx2 (see Sect. 4.2.3) but these were not coupled with cobalt injection.

3.4.3 Response of the Retractor Coxae Motor Neurons to Movement of the Joint (Graham and Wendler 1981a)

A subcoxal joint (middle or hindleg) of restrained animals was moved sinusoidally back and forth at various frequencies. The responses of the five motor neurons described by Graham and Wendler (1981a) are presented in Fig. 3.9 in the form of histograms giving the number of spikes for each motor neuron per 40° of stimulus phase angle (a full stimulus cycle = 360°). It can be seen that the excitatory neurons are active during forward movement of the leg and the CI, during rearward movement.

The responses of these neurons have several parallels to those of the FETi and SETi during stimulation of the femoral chordotonal organ: (1) The slow axon is active over the whole frequency range with a broad maximum between 1 Hz and 5 Hz. (2) The faster axons are active only within a limited stimulus frequency range. The "faster" the axon, the narrower the range. (3) The faster axons fire only during the first few stimulus cycles.

Amputation of the distal portion of the coxa including the rest of the leg, the hair rows on the coxa (see Sect. 6.2.1) as well as the dorsal articulation of the subcoxal joint has little effect on the neural responses. Direct stimulation of the hairplate BF2 or the hair rows on the coxa also produces no clear response. Therefore, the primary sense organ for this reflex must

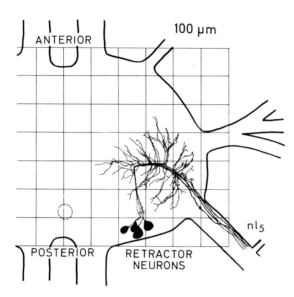

Fig. 3.8. Cobalt chloride back-fill of nerve nl$_5$ showing the retractor motor neurons of a mesothoracic ganglion. *Dotted circle* shows weak filling and probably corresponds to the common inhibitor soma. (Graham and Wendler 1981a)

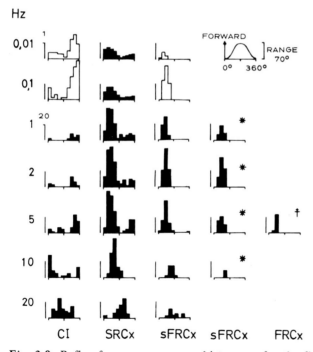

Fig. 3.9. Reflex frequency response histograms for the five axons innervating the retractor coxae muscle during sinusoidal movement of the coxa. *White histograms* have a different ordinate scale. *Asterisks* denote that the response occurs only in the first few cycles of the stimulus; *plus sign,* response does not appear in every cycle of the stimulus. (Graham and Wendler 1981a)

lie between the trochantin and the coxa. Its identity is as yet unknown. The hair rows and the hair plate BF2 appear to be of secondary importance.

If during an experiment the insect makes active movements, the reflex can usually no longer be observed (for reversal of this reflex in walking animals, see Sect. 4.2.6.3). The response to a particular stimulus is high immediately after a disturbance and decreases gradually over time. As in the femur-tibia feedback loop the gain of the system is therefore dependent on the state of activity (see Fig. 2.12).

3.5 Leg Raising by the Standing Animal (C. Walther, pers. comm.)

Tickling the dorsal side of the tarsus of a standing animal with a small brush often causes it to raise this leg and keep it raised for longer than 30 min. If a second leg is similarly stimulated, the insect usually lowers the first leg as it raises the second. With much patience it is possible to get the animal to raise two legs simultaneously. In this case if the tarsus of a third leg is tickled, at least one of the first two legs will be set down again. Figure 3.10 shows which of the two legs is preferentially set down for different combinations of raised legs. Preference was scored when one leg was set down at least three times as often as the other. The results show an order of preference which seems to obey certain rules. The strongest influence is on adjacent ipsilateral legs, whereby the middle leg appears to influence the foreleg more than it does the back leg (Fig. 3.10, 15). Next comes the influence on contralateral legs of other segments, then the influence on the ipsilateral leg of the non-adjacent segment (Fig. 3.10, 10 and 14). The weakest influence is on the contralateral leg of the same segment. If two legs from the same segment are raised and a leg from the non-adjacent segment is stimulated, the contralateral leg is preferentially set down (Fig. 3.10, 3 and 4).

Since the results from experiments on suspended animals (no load on the legs) are nearly the same, the primary effect is probably via nervous channels and not due to the load on the single legs. The load on the legs does, however, play a minor role as shown by the difference between loaded and unloaded legs in situations 14, 18, and 22 (Fig. 3.10).

3.6 Height Control by a Standing Animal

When a stick insect is standing on an uneven surface it holds its longitudinal body axis in different positions according to the shape of the substrate.

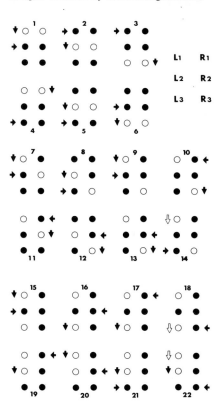

●, leg on ground ➔, stimulated leg
○, leg off ground ▼, reacting leg

⇩ preference only in suspended animals, in
free standing animals no preference

Fig. 3.10. If two legs are raised and the tarsus of a third leg is touched, one of the raised legs will be set down. *Numbers* denote the different combinations of raised and stimulated legs; *downward pointing arrows*, the leg which is preferentially set down. (R. Walther, unpubl.)

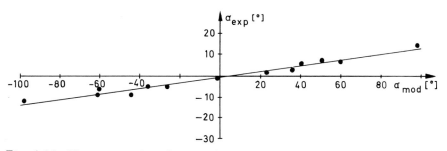

Fig. 3.11. The measured angle α_{exp} between the meso- and metathorax plotted against the angle α_{mod} that would be expected if the vertical distance between each segment and its corresponding tarsi were kept constant, i.e., if the body adjusted to the shape of the substrate by changing the meso-metathoracic angle. The regression line is shown. (Stammer 1978)

The initial investigations of this behavior centered on whether this adjustment to an uneven substrate was accomplished by bending the thoracic joints. Figure 3.11 shows that the angle formed by the meso- and metathorax significantly adapts to irregularities of the substrate. However, in absolute terms this adaptation is so weak that the longitudinal body axis can be regarded in the first approximation as rigid. (The angle between pro- and mesothorax was not investigated because the prothorax is too short to play a significant role.)

If the body is rigid, then the vertical distance between tarsus and body must change markedly depending on the shape of the substrate. This was found to be the case.

There are feedback systems in each of the leg joints. It can thus be assumed that the force exerted by a leg on the ground changes with the vertical distance between tarsus and body (tarsus height). The dependence of this force on tarsus height was measured as follows: The stick insect was placed on three parallel horizontal rods with each pair of legs on a different rod connected to a force transducer. The height of each rod was varied and the resulting tarsus height of each thoracic segment was mea-

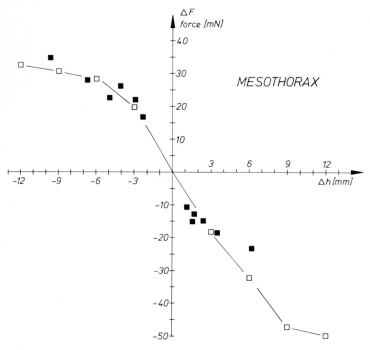

Fig. 3.12. The change in force (ΔF) as a function of stepwise changes in tarsus height (Δh) of both middle legs. The forces were measured immediately after the step. ■ represent unrestrained animals; □ restrained animals. The corresponding curves for fore- and hindlegs are similar. (Stammer 1978)

sured together with the amount of force exerted by the leg. Figure 3.12 shows the results for middle legs of free standing animals. The corresponding curves for fore- and hindlegs are similar.

Experiments in which the body was held stationary, show that the amount of exerted force depends on tarsus height in the same way as in free-standing animals. Furthermore, varying the tarsus height of one pair of legs affects force production only in that pair and not in the other two. The same applies to legs of the same segment when the height of only one leg is changed.

These findings suggest that each pair of legs (apparently even each single leg) possesses its own "height control system," which tries to maintain a certain reference height. The only coupling between the three feedback systems is mechanical via the rigid body of the animal. In these control systems the force is a function of the deviation of the body height from the reference point (given for the middle legs in Fig. 3.12). Model calculations have shown that all of the experimental results are quantitatively compatible with this hypothesis (Stammer 1978; Schmitz 1978; Schreiner 1979).

4 Walking

4.1 Introduction

Walking is the most common mode of locomotion for terrestrial animals. It is highly developed in two systematic groups, in arthropods (especially insects, spiders, and decapod crustaceans) and in four-legged vertebrates (especially mammals). Even a cursory glance reveals differences between the walking systems of most insects and mammals. For example, in insects the legs stand out to the side so that the body rests in a relatively stable equilibrium position. Also, insect legs grip the substrate, furnishing a stable base even when the animal's center of gravity lies outside the area enclosed by the standing tarsi. Thus, at least in intact insects, the control of walking can occur without consideration of balance problems.

Stick insects are nocturnal climbing animals. When this form of locomotion is spontaneous, the leg movement is usually slow and the step frequency, low. When disturbed, stick insects are capable of relatively fast escape runs even in daylight. Until now the study of stick insect locomotion has concentrated mainly on escape runs on a horizontal surface because they are easy to elicit. Even these escape runs are quite slow in comparison to walking of many other insects. Therefore, two things should be kept in mind: (1) In terms of rapidity of movement in insects, stick insects are at the bottom end of the scale. Thus results from experiments on stick insects are not necessarily applicable to fast running insects (e.g., cockroaches). (2) The escape run which has been investigated almost exclusively is not the only mode of locomotion for the animal. The walking system may be better adapted for climbing in the dark than for running away. In spite of these limitations *Carausius* is an ideal object for the study of walking. The following chapters will demonstrate why.

During walking the six legs are moved in a very specific temporal sequence (for details see Sect. 4.4). For example, the ipsilateral fore- and hindlegs are swung forward simultaneously or almost simultaneously. If the middle legs are amputated, the ipsilateral fore- and hindlegs move in antiphase, i.e., with a totally different coordination (v. Buddenbrock 1921). This demonstrates that the movements of all the legs are not controlled by a single inflexible "master center" but that each leg has its own control center which is somehow coordinated with the control centers of the other legs.

From this, it follows that the study of the control of walking can be subdivided into two aspects, "control of a single leg," and "mutual influences between the legs". Mutual influences can determine the timing of leg movements so that movements of each leg occur in a definite temporal sequence (coordination). They can also affect the way a leg is moved within this temporal frame. Thus, this chapter has three subdivisions: (1) control of a single leg, (2) mutual influences between the legs that do not influence timing, and (3) control of leg coordination.

4.2 Control of a Single Leg

This and the following chapters deal solely with normal forwards walking. Backwards walking is discussed in Section 4.5. The animals were touched on the abdomen to induce walking and sometimes to prolong it under daylight conditions.

The treatment of control mechanisms for the movement of a single leg is based on the following premises: There exists an underlying *program* for leg movement. Like a computer program it contains a set of "instructions" for directing basic motor neuron operations and/or appropriate responses to particular afferent input and/or to information from other parts of the CNS. The program is laid down in the properties and synaptic connections of the participating neurons. It is not identical with the temporal sequence of the activities of single motor neurons (the term "program" is often defined this way), rather it is the "instruction" or better: the cause for the production of these activities in their characteristic sequence (see also Sect. 2.5.4). The intensity and temporal arrangement of motor neuron activity is referred to here as the *motor output.*

A program for the control of a leg during walking could be organized in three alternative ways. (1) It could be designed to operate without any feedback, i.e., sensory messages would have no essential effect on the motor output "walking." The entire underlying neural network would then perforce be localized within the CNS. Such a program is called here a *central program.* (2) It could be a *central-peripheral program* in which feedback is not absolutely necessary for the production of "walking" but would under normal circumstances, i.e., intact sense organs, essentially determine the motor output. The oscillator that controls rocking is an example of this type. (3) Finally sensory messages could comprise an essential component of the program, without which the motor output "walking" could not be produced. This would be a purely *peripheral program.*

The boundaries between these alternatives are not sharp and depend on the definition of "essential." Is a sensory message which determines the frequency of an oscillator "nonessential" merely because the program alone can produce a rhythmic output (albeit with a different frequency)? According to the definitions in Section 2.5.4 and Fig. 2.21, a program is central-peripheral if eliminating the periphery changes motor output characteristics (like frequency, intensity, etc.). As can be seen from the literature, there is no universal consensus on the definition of "essential" and therefore of central-peripheral.

The above classification is perhaps too schematic. However, in view of the significance which the central versus peripheral controversy has had in the past, this distinction is maintained (see also Sects. 2.5.4 and 4.2.8).

4.2.1 Description of the Walking Movement of a Single Leg

4.2.1.1 Free-Moving Animals

It is immediately apparent that a step consists of two parts: (1) the stance phase (retraction) in which the tarsus remains on the ground, stationary relative to the substrate while the leg moves backwards relative to the body and (2) the swing phase (protraction), in which the leg is swung forward through the air. Figure 4.1 illustrates the motion of a foreleg during a stance phase of particularly large amplitude such as is often

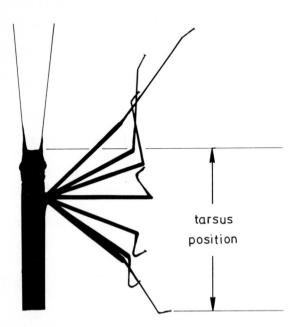

tarsus

position

Fig. 4.1. The movement of a right foreleg of an adult stick insect with amputated middle legs was filmed during the stance phase at 18 frames/s and every other frame was traced. The tarsus position is indicated for the posterior extreme position

observed after amputation of the middle leg. Figure 4.2 shows the positions of the tarsi of all legs of an intact animal at various moments during a walk. The relatively equal spacing between the open circles indicates that swing phase proceeds at an approximately constant speed. The protraction duration is 100 ms–150 ms (Graham 1972; Cruse and Saxler 1980a). The movement is quite slow at the beginning and end of the stance phase. During the middle part of retraction the speed is faster and almost constant (Bässler 1972a; Cruse 1976a).

When decerebrate animals walk slowly, the swing phase may be interrupted by pauses especially in foreleg movement. In the extreme case such a swing phase might proceed like this: raise leg, pause, protraction, pause, swing down, tarsal contact (Graham 1979b). The stance phase may also consist of two parts (see Sect. 4.2.3).

Before the beginning of the stance phase the tarsus grips the substrate. Before the beginning of the swing phase it releases its hold on the substrate. Thus, there must be a gripping phase between swing and stance phase and a release phase between stance and swing phase.

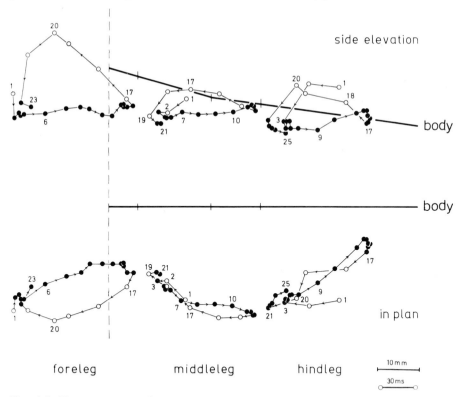

Fig. 4.2. The movement of the tarsi relative to the body of the insect when it is walking on a horizontal plane, as seen from above and the side. ● denote when the tarsus touches the ground. (Cruse 1976a)

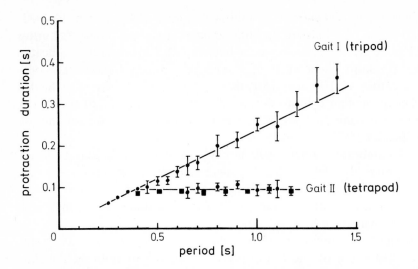

Fig. 4.3. Protraction duration t_p as a function of period P for the mesothoracic legs of first instar and adult insects. ● give average values of t_p for a first instar insect; ■ average values of t_p for an adult insect; *error bars* the error of the mean where it is larger than the symbol. (Graham 1972)

The duration of a step can vary greatly. The highest observed step frequency in free walking adults is 3 Hz (can be somewhat higher in larvae). Changes in the total duration of a step have very different effects on the individual phases. In the normal tetrapod walk of the adult (for definition of different gaits see Sect. 4.4) and in the tetrapod walk of the first instar, swing phase duration is independent of step duration (Wendler 1964; Graham 1972). On the other hand in the tripod gait (see Sect. 4.4.1) of first instar larvae swing phase duration increases with increasing step duration (Fig. 4.3; Graham 1972), whereby approximate proportionality persists between total step and protraction duration. There is also an approximate proportionality between step and protraction duration in decerebrate walking. The slope of the curve is about halfway between those for tripod and tetrapod gaits (Graham 1979b).

The tarsus position was used to describe leg position as a single quantity (for justification, see Bässler 1972a). It was defined as the distance between the tarsal tip and an imaginary line drawn through the tip of the head perpendicular to the longitudinal body axis (see Fig. 4.1). The anterior and posterior extreme positions are dependent on external conditions (see Fig. 4.4). Sometimes these extreme positions change in the same direction (e.g., loaded animal) which tends to keep the step length constant. In other cases, the step length also changes (e.g., walking uphill).

Step length is almost independent of walking speed for the tripod straight walk of first instar larvae. During a direction change, step length

horizontal   up the vertical   horizontal plane
path   path   with load

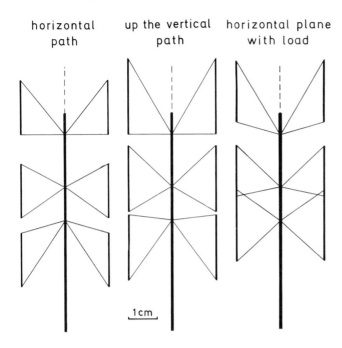

Fig. 4.4. The anterior and posterior extreme positions (for tarsus position see Fig. 4.1) for the walking situations: horizontal path, up a vertical path, and horizontal plane with backward-directed load. (Cruse 1976a; Bässler 1977a)

decreases on the inside of the turn (Graham 1972). Turning third instar larvae not only decrease step length on the inside (Jander and Wendler 1978) but also increase it on the outside of the curve (Jander 1978). In the adult insect stride length changes only slightly during turning (Graham 1972), and turning is accomplished by varying step frequency on both sides of the body. As in the larvae step length is only slightly dependent on walking speed (Graham 1972).

4.2.1.2 Adults Walking on a Treadwheel

Walking studies are often easier to carry out on an animal fixed above a treadwheel than on an unrestrained animal. Wendler (1964) and Graham and Wendler (1981b) used a single, relatively heavy wheel. The insect was supported a set distance above it and leg movements were recorded automatically. As a further development Graham (1981) designed a treadwheel constructed of two very light wheels which share a common axle but which can be moved independently of each other. The animal is fastened above it so that the right legs walk on one wheel and the left legs, on the other. The common axle is counterbalanced so that the wheel exerts an upthrusting force about equal to the insect's body weight. The animal is

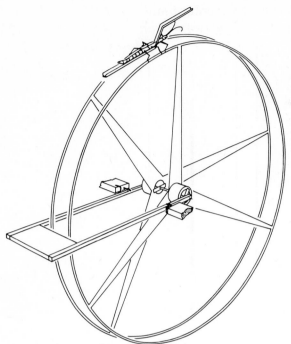

Fig. 4.5. Double treadwheel (Graham 1981a). The moment of inertia of each of the two wheels is 300 g · cm², corresponding to an effective inertia of about twice the body weight of the animal

fixed to a support rod and can set the distance of the wheels from its body (Fig. 4.5). Both wheels are turned by the insects themselves.

Movements executed on such a double treadwheel correspond on the whole to those of a free walking animal. One notable exception is that on the lightest wheels the protraction duration depends on the walking speed. This is not the case for free walking adults. At the same time the animals use the tripod gait (Graham 1981). If a very lightweight double wheel with very low friction is used, protraction duration also increases with increasing period duration (but less steeply than during the tripod walk shown in Fig. 4.3). However, when the friction of the same wheel is increased, protraction duration becomes independent of period duration and is approximately the same as in free walking adults. Also, both the average and minimum step durations increase. When friction is applied to the wheel, the animals exhibit a tetrapod gait; with no applied friction, the gait is tripod (Foth and Graham, unpubl.). Thus, the durations of stance and swing phase depend on environmental parameters. In free walking adults the abdomen dragging along the ground probably produces enough friction to bring the animal under conditions similar to those of the treadwheel with applied friction.

4.2.1.3 Animals on a Mercury Substrate

Legs that are on the ground at the same time interact mechanically with each other during walking due to substrate coupling. A double tread-wheel system like Graham's (1981) eliminates only the mechanical coupling between left and right sides. In order to study the way a leg moves completely independently of mechanical influences, the walking of adult insects on a mercury substrate was studied (Graham and Cruse 1981). The insects were supported at normal body height over the mercury. The viscosity of mercury is so low that the legs do not influence each other mechanically, whereas its surface tension is high enough so that the legs do not sink in.

Leg movements are fully coordinated under these conditions. Step frequency increases to a maximum of 5 Hz which is significantly higher than for free walking or wheel walking animals. Mean swing phase duration is only slightly prolonged (Fig. 4.6). A plot of swing phase duration as a function of period duration shows that swing phase duration increases with step duration (exactly like nymph free walking and similar to adult walking on a low friction treadwheel). Since the resistance that the leg has to overcome during the stance phase is less on mercury than in normal walking, the stance phase duration must be determined by the time which the muscles need to bring the leg all the way back to the posterior extreme position.

Leg movement is not only faster, it is also usually interrupted by pauses of about 50 ms between stance and swing phase and between swing and stance phase. It appears as if both swing and stance phase have to be initiated by afferences.

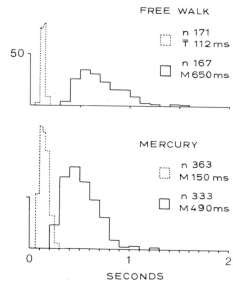

FREE WALK

n 171
$\overline{T}$ 112 ms

n 167
M 650 ms

50

MERCURY

n 363
M 150 ms

n 333
M 490 ms

0   1   2

SECONDS

Fig. 4.6. Distribution of period duration *(solid lines)* and swing phase duration *(dotted lines)* of adults walking free and supported above mercury. *n* no of measurements; *T* mean; *M* median. (Graham and Cruse 1981)

4.2.2 Description of the Torques in Single Leg Joints
During Stance Phase (Cruse 1976a)

The intensity and direction of the force exerted by a single leg on the substrate during stance phase was measured using force transducers oriented in each of the three spatial dimensions (Fig. 4.7). If the positions of all joints are known for each point in time, the torques affecting each joint can be calculated using static methods. Besides a precise description of the time course of torques, the details of which can be found in the original publication, these experiments yielded the following results of general interest:

1. The torque can oppose the observed movement. When the insect is walking on a horizontal surface, for instance, it extends the femur-tibia joint of the hindleg in the second part of the stance phase. At the same time a "flexing" torque is registered in this joint.
2. The timing and amount of the torque of a particular joint is different in different walking situations (on a horizontal surface, on or hanging from a horizontal bar, on a vertical bar): on the horizontal surface the forelegs have a detector function, the middle legs carry the weight of the body (they produce a force directed anteriorly in the first part of the stance phase and posteriorly in the second part), and the hindlegs provide addi-

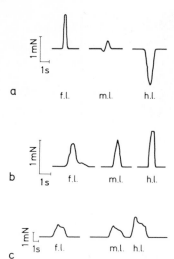

Fig. 4.7a–c. The time course of the force components in the **a** longitudinal (backwards = positive), **b** transverse (against the body = positive) and **c** vertical axis (downwards = positive) when the animal is hanging from a horizontal bar. The direction and intensity of the force exerted by each leg on the substrate can be read from the curves for a particular point in time. If the position of the individual leg joints is known for this time, the torque can be calculated for each joint. *f.l.* foreleg; *m.l.* middle leg; *h.l.* hindleg. (Cruse 1976a)

tional support and a continuous backward directed thrust. The animal uses all its legs to carry the body when hanging beneath a horizontal bar. Forward thrust comes mainly from the middle legs. In a vertically upwards walk all legs provide support and forward thrust. These experiments show that the motor output can be greatly influenced by sensory input.

The torque exerted on a joint is generally produced by all antagonistic muscles. The decrease for example of a "flexing" torque can be produced by decreased activity in the flexor or increased activity in the extensor. This method cannot distinguish which of these two possibilities actually occurs. Thus, the beginning and end of activity of a particular muscle cannot be determined from such measurements.

4.2.3 Motor Neuron Activity During a Normal Step

Recordings from the motor axons of the extensor tibiae muscle can be made in *free walking animals* using insulated wire electrodes inserted through the cuticle (see Sect. 2.8.2). With this method only myograms are obtainable from the other muscles. In muscles with complex innervation patterns myograms only show the beginning, end, and mean activity of the excitatory units.

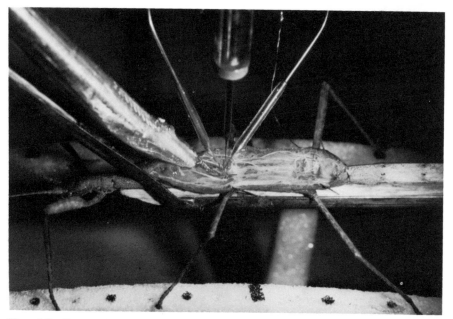

Fig. 4.8. Set-up for intracellular recording from treadwheel-walking animals. The thorax is opened dorsally and the gut removed to expose the ventral nerve cord for electrode insertion. (Graham and Godden, in prep.)

Nerve recordings can also be obtained from animals *walking on a tread-wheel*. The thorax is opened dorsally and removal of the gut exposes the ventral nerve cord and its branches for *extracellular* recording (Figs. 4.8 and 4.9, lower trace). Animals prepared in this way walk like intact animals. At the beginning of an experiment behavioral tests are carried out (see Fig. 3.9) which together with relative spike height makes axon identi-

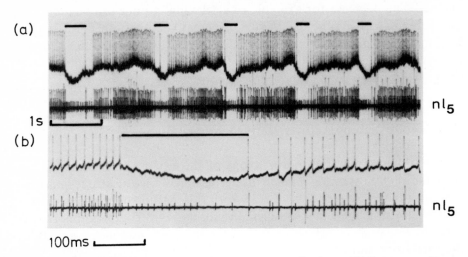

Fig. 4.9a, b. Intracellular recording from the neuropil of an sFRCx neuron (*upper traces:* **a, b**) and simultaneous recording from nl$_5$ (*lower traces:* **a, b**) from an animal on a treadwheel. The time scales of **a** and **b** differ. Protractions of the corresponding leg are designated by the *black bars* at the top. (Graham and Godden, in prep.)

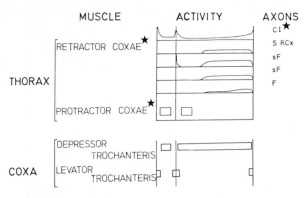

Fig. 4.10. Electrical activity of single identified motor neurons and whole muscles of a middle leg during a normal step. Only those muscles which have been studied are shown. Neuronal spike frequency is given from 0–200 Hz as a function of the phase. The activity of whole muscles is represented by *blocks*. *Star* denotes muscles that are innervated by the common inhibitor. (According to Graham)

fication in such recordings possible. This has been done for all motor neurons which have so far been identified (especially the motor neurons of the retractor coxae muscle, see Sect. 3.4.2).

The ganglion can be totally immobilized when an animal is walking on a treadwheel. Therefore, *intracellular* recordings can be obtained (Graham and Godden, in prep.). As yet, this has only been successful for the motor neurons of the retractor coxae muscle (see Fig. 4.9).

A summary of results (combined from myograms, extra- and intracellular recordings) for the middle leg is shown schematically in Fig. 4.10. In this leg the subcoxal and the coxa-trochanter joints are the primary joints moved during walking. Therefore, analysis has until now been limited to the muscles of these joints. In two cases activity was recorded that was not expected from the leg position.

1. If the leg touches the ground at the end of a swing phase, there is a short burst of activity from the levator trochanteris muscle. This may assist gripping.
2. At the beginning of a stance phase the protractor coxae muscle often shows intense activity. The first phase of retraction is thus passive (the leg is moved backwards due to mechanical coupling with the other legs). There is even a resistance opposing this passive movement (see also Sects. 4.2.1.3 and 4.2.2). Apparently the protractor activity at the beginning of the stance phase is a resistance reflex.

4.2.4   Afferent Influences on the Walking Movement of a Single Leg

Usually the first question to be answered about the control mechanisms of a particular behavior is whether its basis is central or peripheral, i.e., what are the relative contributions of CNS and periphery. Since the motor output accommodates to changed conditions (see Sects. 4.2.1 and 4.2.2), peripheral sensory influences must be a part of the walking program. The next step in the investigation of this system was to locate this part of the program.

Ablation of some of the sense organs does not drastically alter walking (Bässler 1973, 1977b). For example, after removal of the hair plate BF1 on the coxa-trochanter joint (see Sect. 6.2.2), the leg swings out higher during protraction. After the hairs of BF2 on the subcoxal joint (see Sect. 6.2.1) have been shaved off, the leg swings farther forward. These changes are quantitatively significant but do not in principle alter the movement form (Wendler 1964, 1965a). On the other hand, distinct qualitative changes in the leg movement do occur if certain sense organs of a leg are manipulated in such a way that they produce an *"experimentally induced" afference.* Such an afference is defined as a sensory input mainly generated by the experimental situation and not by the motor output of the animal. This afference can either be incorrect, for example, when it signals a joint position which does not coincide with reality or it can be correct as when it reports an experimentally imposed position.

4.2.4.1 Femoral Chordotonal Organ

In order to experimentally induce an incorrect afference, the receptor apodeme is removed from its insertion on the tibia, dorsal to the joint's rotational axis and attached to the tendon of the flexor tibiae muscle *("crossing the receptor apodeme").* Since the flexor tibiae tendon inserts ventrally to the rotational axis of the femur-tibia joint, the chordotonal organ now reports the opposite of the actual joint movement. In these experiments the average extension of the chordotonal organ is generally higher than in intact animals. A small part of the flexor tibiae muscle must be removed in this operation (for details, see Bässler 1967).

Any leg with crossed receptor apodeme exhibits a very characteristic behavior during walking. The leg is carried quite high with the femur-tibia joint fully extended (Fig. 4.11) and the basal joints making small amplitude movements (twitches). For obvious reasons this behavior has been termed "saluting."

Fig. 4.11. Stick insect saluting with its left middle leg in which the receptor apodeme has been crossed

Upon tarsal contact (either due to basal joint movement bringing the tarsus into contact with the substrate or because an object is held against the tarsus) the tarsus grips the substrate and a stance phase is executed in which the femur-tibia joint is usually flexed. When retraction is completed, the leg is raised in the posterior extreme position. This may be either followed by a normal swing phase and even several full steps or the leg is again swung forward but kept high in the air more or less motionless. Many animals do not execute any stance phases on a level surface whereas some insert one or more stance phases at irregular intervals. Still others take almost as many steps with the operated leg as with the intact ones.

An operated hindleg that has reached its posterior extreme position at the end of a stance phase often maintains its grip on the substrate instead of rising into the air. When this happens, it is usually visibly jerked from the ground by the activity of the other legs. This type of gripping has been observed only in hindlegs (Bässler 1967, 1977b).

The responses produced by crossing the receptor apodeme do not commence immediately. Rather the positions are reached via the execution of a more-or-less normal step. The legs appear to "freeze" in a particular part of their normal movement cycle, the swing phase. Furthermore, the hindlegs may not carry through a complete stance phase.

How can these observations be explained? Normally the femur-tibia joint of an intact leg is extended during a swing phase. At the end of a stance phase it is also extended in the hindleg. When the receptor apodeme is crossed, the afferents from the chordotonal organ incorrectly signal that the joint is flexed when it is actually extended, i.e., that the phase in question is still being executed. The stance phase begins when tarsal contact is reported. Tarsal contact normally signals the end of swing phase. This leads to the testable hypothesis that when sense organs report that a certain phase of leg movement has not yet been completed, the leg either remains in this phase or only begins the next phase after a delay.

4.2.4.2 Campaniform Sensilla on the Trochanter

After the leg has released the substrate at the end of stance phase, it is no longer under tension. A signal from sense organs (e.g., campaniform sensilla) that the leg is still under tension, would mean that the process of releasing the substrate was not yet completed. In order to investigate the influence of the campaniform sensilla on walking, the trochanter groups (see Sect. 6.2.2) were continuously stimulated in free walking animals by attaching a steel wire clip to the trochanter. Probably sense organs besides the campaniform sensilla are also stimulated (above all tactile hairs).

In spite of this very unspecific mode of stimulation the leg is no longer swung forward during walking. It is, however, swung forward during

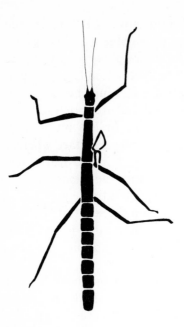

Fig. 4.12. Typical posture of a right middle leg during walking after a wire clip has been attached to its trochanter

searching movements demonstrating that the clip does not mechanically hamper leg movement. During walking the leg is held as shown in Fig. 4.12 and dragged over the ground. The tarsus is raised and the femur-tibia joint is usually immobile. If the ventral side of the tarsus touches an object, the tarsus does not grip it. In other experiments the clip is attached to the leg of a standing inactive animal and the animal is induced to walk by touching its abdomen. During the first stance phase, the clipped leg moves normally with the tarsus gripping the ground. After this leg reaches its posterior extreme position, the insect raises its tarsus but does not swing the leg forward, dragging it along while walking further. No difference between fore-, middle and hindlegs have been observed (Bässler 1977b). The results of these experiments are in agreement with the hypothesis formulated for the crossed receptor apodeme. The response (dragging along of the leg with raised tarsus) does not occur immediately but is brought about by the execution of a step. The leg remains stuck in that phase of the movement which the experimentally-induced afferences signal as incomplete.

The fact that continuous stimulation of the campaniform sensilla can suppress leg protraction during walking but not during searching movements, is a further example of a *program-dependent reaction* (see Sect. 3.2.3).

4.2.4.3 Position Receptors on the Subcoxal Joint

When a wooden rod is presented to a treadwheel-walking stick insect, it grasps the rod with one or two legs and does not let go as long as its other legs are turning the treadwheel (experimental set-up as in Fig. 4.14). If the wooden rod is moved backwards, the leg is usually lifted from it (end of stance phase) only when a position approximately corresponding to the posterior extreme position is reached. The leg produces a relatively large, backward-directed force as long as it is gripping the rod (Wendler 1964). Apparently the leg remains in the stance phase until its proprioceptors signal the end of this phase. In the hindleg the femoral chordotonal organ is probably one of these proprioceptors (see Sect. 4.2.4.1). Due to the way the fore- and middle legs move, it seems likely that the posterior extreme position is primarily signalled by the sense organs monitoring the position of the subcoxal joint. To test this, the hair rows were removed from one leg, and warm plasticine was applied to the hair plate BF2 on its subcoxal joint so that these hairs were bent away from the joint. This procedure served to deliver continuous stimulation from hair rows and BF2 corresponding to that which is normally encountered in the anterior extreme position. The internal receptors on the trochantin were left intact (see Sect. 3.4.3).

The posterior extreme position of the operated leg shifts rearwards, often as far as is anatomically possible. This leads to conspicuously unnatural leg positions (Fig. 4.13). During slow walking the operated leg

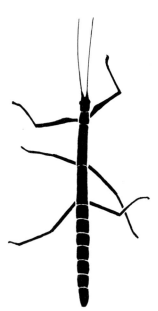

Fig. 4.13. Typical walking position of a right middle leg from which the hair rows on the subcoxal joint have been removed and in which the hairs on the hair plates have been fixed in a bent position

occasionally does not release the substrate when it has reached the posterior extreme position. The operated leg is then torn with a distinct jerk
from the ground by the activity of the other legs. This behavior, gripping
followed by tearing loose by the other legs, was observed in fore- and
middle legs but never in hindlegs and is identical to that of hindlegs with
crossed receptor apodemes at the end of stance phase. Normally, especially during fast walking, all legs are lifted as usual but always in the experi-
mentally-shifted posterior extreme position. The anterior extreme position is also shifted rearwards (Bässler 1977b).

Apparently the transition from swing to stance phase is also influenced
by proprioceptors on the subcoxal joint. Wendler (personal communica-
tion) presented brief electrical stimuli to the retractor coxae muscle of a
leg during swing phase. This prevents the leg from moving forward during
swing phase. After the end of stimulation the leg first moves forward to
its normal anterior extreme position and only then starts a normal stance
phase.

These findings permit the following generalizations: (1) All observable
responses to experimentally-induced afferences resemble leg positions
which occur under natural conditions. The surgical procedures seem only
to "exaggerate" them. (2) The stimulus never elicits a direct and immediate response, rather the leg assumes this position via a more-or-less normal
step. It then remains in this position for a period of time if not perma-
nently. (3) The stimuli that cause the leg to linger in a particular phase
are stimuli which signal that the phase in question is not yet completed.
(4) A leg's normal movement cycle can be interrupted in at least three different places: at the end of stance phase (before release of the substrate),
after release of the substrate (before the start of swing phase), and during
swing phase. Since simple ablation of the respective sense organs does not
have much of an effect, the end of one phase and the beginning of the
next must also be influenced by other channels.

4.2.5 Motor Output During Prolongation of a Phase

As has been shown, afferences that signal that a particular phase of the
step cycle has not ended, can keep the leg in this phase. What is the motor
output of a leg that remains "stuck" in a certain phase? This question has
been investigated in two situations, remaining in stance phase and remaining in swing phase.

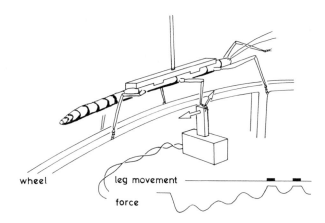

Fig. 4.14. A leg that is on a force transducer to the side of a treadwheel exerts a backwards-directed force which is rhythmically modulated. It is occasionally raised briefly during a force minimum

4.2.5.1 Remaining in Stance Phase

In these experiments the animal was fastened over a treadwheel so that one leg stood on a force transducer (see Fig. 4.14). When the other legs are walking, this leg exerts a backwards-directed force on the force transducer (Wendler 1964; Bässler 1967). The force is modulated in the stepping rhythm of the other legs and coordinated with them (Cruse and Saxler 1980a, b).

Studies on first instar larvae of *Extatosoma* (Bässler 1979) have shown that the leg on the force transducer is actually in the walking state. These animals walk very quickly on the treadwheel (up to 5 steps/s) and with regular steps. The behavior of the leg on the force transducer shows a continuous transition to normal walking: a continuous rhythmically modulated, backwards-directed force (as is normally the case in *Carausius*); brief rhythmic lifting of the leg during the force minima (as illustrated in Fig. 4.14.); longer lifting with replacement on the force transducer; and leg lifting followed by the execution of a normal step on the treadwheel.

When the treadwheel is stopped, the force oscillations persist only if at least one of the other legs is still moving. This indicates that the source of these oscillations is probably to be found outside of the system responsible for control of the leg on the force transducer (Cruse and Saxler 1980a).

The electrical activity of metathoracic legs on the force transducer has been recorded in the retractor and protractor coxae (nerve recordings, unpublished) and from the flexor (myogram) and extensor (nerve recording) of the tibia (Schmitz 1980). Generally only the retractor is active in the protractor-retractor system. Its activity is usually quite high and is modulated in the step rhythm of the other legs (as in Fig. 4.21). In the flexor-extensor system of the hindleg this kind of regular modulation is not apparent.

4.2.5.2 Remaining in Swing Phase (Saluting)

Myograms from the protractor and retractor coxae muscles and the levator and depressor trochanteris muscles were recorded from animals with crossed receptor apodemes (see Sect. 4.2.4.1) walking on a double treadwheel (Graham and Bässler 1981). Only middle legs were investigated. Figure 4.15 shows an original recording, and Fig. 4.16, a summary of the results. When the operated leg is performing a stance phase on the treadwheel, the motor output to all four muscles is normal. During saluting (see Sect. 4.2.4.1) the protractor and levator muscles are most active. This produces a widely swung-forward coxa and a highly raised femur. As in the other two muscles, their activity is rhythmically modulated with a frequency of about 3–4 Hz. The other legs in these recordings walked at a frequency of about 1 Hz. The modulation frequency is, thus, distinctly higher than the walking frequency and about the same as searching move-

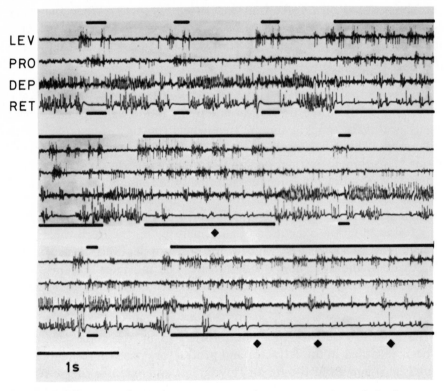

Fig. 4.15. Myograms from the protractor and retractor coxae and the levator and depressor trochanteris muscles of a right middle leg with crossed receptor apodeme. *Black bars* indicate when the leg is not in contact with the treadwheel, i.e., when it is in a normal swing phase or a salute. *Black diamonds* mark the onset of a resting period after a significant forward swing. (Graham and Bässler 1981)

ment frequency during swing phase (Sect. 4.2.7.3). The high frequency oscillation does not produce any clearly visible oscillation in leg position (Fig. 4.16). Some of the activity bursts are stronger than others (especially in the depressor), producing brief anteriorly-directed movements (twitches). These occur in rhythm with the other legs and are coordinated with them (Fig. 4.16). The motor output of a middle leg during a salute is, thus, modulated with two superimposed frequencies. The higher frequency oscillation affects all the muscles studied. Often the levator and protractor muscles are simultaneously active, as are the depressor and retractor muscles. It is possible that these oscillations are weak searching movements since they lie within the same frequency range as some searching movements of intact legs. The lower frequency oscillation corresponds to the walking frequency of the other legs. It preferentially effects the depressor muscle.

Two observations support the interpretation that "saluting" represents an exaggerated and very prolonged swing phase: (1) A salute starts with the same phase relationships relative to the other legs as does a normal swing phase. (2) At the beginning of a salute the protractor muscle starts to fire after the levator, exactly as in a normal swing phase (Fig. 4.15). These two muscles become synchronized only during the course of a salute.

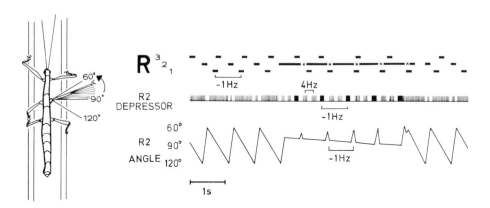

Fig. 4.16. *Left* the animal is shown fixed over two independent wheels. The receptor apodeme of the right middle leg is crossed. *Top right* the step pattern of the legs on the right side during normal walking and saluting. *Short black bars* represent a swing phase; *long bars* a salute. The crosses during saluting symbolize the occurrence of a visible movement of this leg (see *bottom trace*). *Middle right* schematic representation of the electrical activity of the depressor muscle of the right middle leg during the walk shown in the top trace. *Lower right* the angle formed by the femur of the middle leg and the longitudinal body axis during the same walk. The time scale is *from left to right*. (Modified from Graham and Bässler)

4.2.5.3 Conclusions

Motor output is rhythmically modulated during the prolongation of a phase of leg movement. This modulation has the same frequency as the movements of other walking legs. It may be the expression of a rhythmically modulated component within the CNS (oscillating component). This oscillating component does not affect all active muscles in the cases which have been studied so far (middle leg during swing phase and hindleg during stance phase).

It is not yet clear how this oscillating component arises. The following possibilities were considered:

1. The walking program is a central-peripheral program. Each leg has its own central rhythm generator. The oscillating component reflects the activity of the rhythm generator (a central oscillator according to the definition given in Sect. 2.5.4). Since the oscillating component is coordinated with the movement of the other legs, the execution of this program must be modifiable by the other legs.

2. The walking program is a peripheral program, i.e., no central rhythm generator for walking exists. Movement of a leg is controlled exclusively by its own sense organs. The slightly modulated excitation of the sense organs of a leg which "gets stuck" in a particular phase, suffices to drive the program. As in the first case the oscillating component would express the activity of a rhythm generator, but a peripheral one in this case. Here, too, there must be coordinating influences from the other legs.

3. Neither of the above possibilities occurs, instead the program itself remains "stuck" in a particular phase. In this case the oscillating component would express the coordinating influences from other legs or influences from a higher control center.

The following chapter describes experiments which make it possible to decide which of these possibilities actually applies.

4.2.6 Structure and Localization of the Walking Program

4.2.6.1 Ablation of the Periphery

All the experiments discussed in this section were performed on animals walking on a single treadwheel which could not be moved up or down. The body was opened dorsally. Two recording electrodes were inserted, one onto nerve nl_2 which innervates the protractor coxae muscle, the other onto nl_5 which innervates the retractor coxae muscle. In addition either the activity of nl_2 or nl_5 from a neighboring leg or the movement of this leg was monitored. The ganglion half that was recorded from was denervated, i.e., nerves na, nl_1, nl_3, nl_4, and np were cut close to the ganglion; nl_2 and nl_5, distal to the recording electrodes; and ncr, directly after nl_5 branches off (cf. Fig. 6.6).

In the following discussion the term, activity, is applied only to the excitatory motor neurons and not to the common inhibitor (see Sects. 3.4.2 and 6.3.1) whose spikes can also be recorded in this preparation. When the other legs are walking, there is a rhythmic activity alternating between the protractor and retractor motor neurons. The rhythm is exactly the same as the step frequency of the other legs. The difference between maximal and minimal activity in a nerve (expressed not only as the number of active motor neurons but also as their spike frequency) is less than in an intact walking leg. Usually there is some activity in at least one of the two nerves even during the activity minimum (Fig. 4.17). These findings apply to the meso- and metathoracic ganglia. The prothoracic ganglion has not yet been investigated.

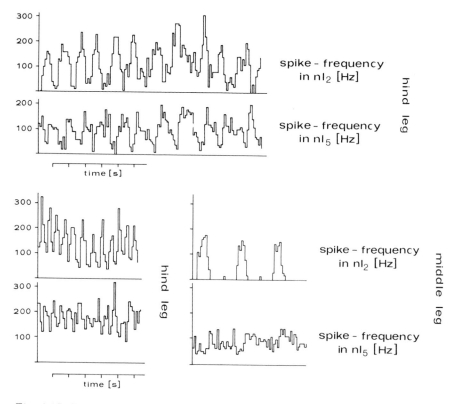

Fig. 4.17. Three typical examples of the activity of the excitatory protractor (nl_2) and retractor (nl_5) motor neurons recorded from a denervated ganglion half while the other legs are walking on a treadwheel. The modulation frequencies correspond to the step frequencies of the walking legs

Since the oscillating component is still present even after denervation, possibility (2) from Section 4.2.5.3 can be eliminated. The oscillation must either express the activity of a central rhythm generator or originate outside of the ganglion half investigated.

In a further set of experiments recordings were made from denervated meso- or metathoracic ganglia while all the other ganglion halves were denervated, one after the other, by cutting all nerves near the ganglion. The regular rhythm persists as long as one leg is intact and can walk on the treadwheel. The rhythm is always the same as the step frequency of the walking leg or legs (Bässler and Wegner, in prep.). In all these experiments the remaining leg was a foreleg.

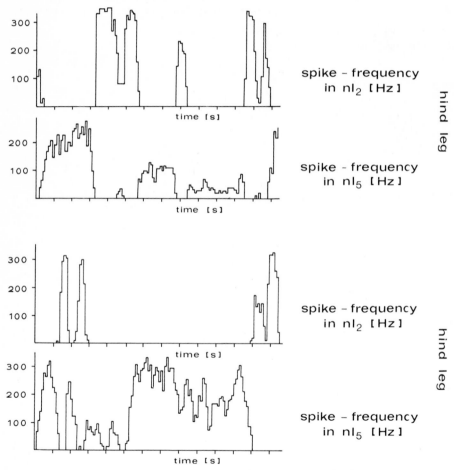

Fig. 4.18. The activity of the protractor (nl_2) and retractor (nl_5) motor neurons recorded from preparations in which the thoracic ventral nerve cord was totally denervated except for the connections to subesophageal and abdominal nervous systems. The animals were stimulated mechanically on the abdomen

When the whole thoracic ventral nerve cord is denervated (leaving the subesophageal and abdominal nervous system intact), there is normally no longer any regular alternating rhythm (Fig. 4.18). Instead, very strong activity appears in either protractor or retractor motor neurons when the animal is stimulated on its abdomen.

To determine how far this motor output corresponds to walking motor activity, the experiments (with completely denervated thoracic ventral nerve cord) were extended to include: (1) The activity of protractor and retractor motor neurons from two neighboring ganglion halves were recorded simultaneously in order to determine whether there was any coordination between these neurons. (2) In one ganglion half the activity of nl_3 in which the FETi and SETi are clearly recognizable was recorded in addition to the activity of protractor and retractor neurons. Out of a total time of 1 h and 40 min for simultaneous recording of protractor and retractor activity in one leg (22 middle and 26 hindlegs), only about 15 minutes of recording fit the criterion for regular alternation. Regular alternation was defined as the sequential appearance of at least two alternating cycles with approximately equal burst durations. Figure 4.19 shows examples of regular alternating activity (about 5 s at the end of the middle leg recording; about 5 s in the first half of the hindleg recording, and another 3 s at its end).

The majority of periods showing irregular alternation between protractor and retractor activity can hardly be regarded as walking activity. The following arguments demonstrate why even the few periods in which

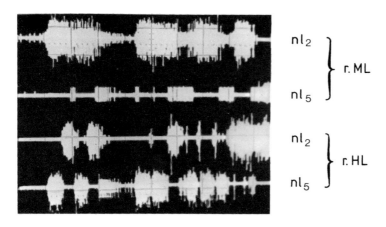

1s

Fig. 4.19. Recordings from protractor (nl_2) and retractor (nl_5) motor neurons of the right middle and hindlegs of a denervated ventral nerve cord preparation. The alternations are unusually regular. The animals were stimulated mechanically on the abdomen

protractor and retractor activity alternate regularly cannot be considered
to be walking.

1. Even during these regular segments the protractor and retractor activity in adjacent legs is not coordinated. Regular alternation in one leg can
occur at the same time as irregular alternation in an adjacent leg (Fig.
4.19, first half). Also, simultaneous regular alternation in two adjacent
legs can have completely different frequencies of alternation (Fig. 4.19,
second half).

2. Comparison of the activity of individual retractor coxae motor neurons
during regular alternation of protractor and retractor neurons and during a
normal step reveals distinct differences (cf. Fig. 4.10). In contrast to that
of a normal step, the firing frequency of the common inhibitor is only
weakly modulated.

3. During many sequences of regular alternation between protractor and
retractor activities the FETi and SETi are continuously active. In the cases
where their frequency is also modulated, there is no clear coordination
with the alternation of protractor and retractor activity.

4. In another set of experiments the whole thoracic ventral nerve cord was
denervated except for the nervus cruris of one leg, which was left intact
up to the first third of the femur leaving the muscles and sense organs of
the coxa and the femoral chordotonal organ innervated. Nerves nl_3, F121,
F122, F2, F3, and F4 were cut. Recordings were made from the stumps
of nl_2, nl_3, and nl_5. The motor output in nl_2 (protractor) and nl_5 (retrac-

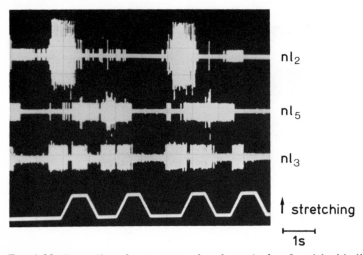

Fig. 4.20. Recordings from nerves nl_2, nl_5, and nl_3 of a right hindleg in a denervated
thorax preparation in which only the nervus cruris serving this leg was intact up to the
beginning of the femur. *Bottom trace* shows stimulation of the femoral chordotonal
organ of this leg during strong activity in all three nerves

tor) is identical to that of the completely denervated ganglion. Stretching the femoral chordotonal organ during regular or irregular alternation of protractor and retractor activity either produces no response in the FETi and SETi or inhibits their activity for a short time (Fig. 4.20). In a resting or walking intact animal this stimulus produces an excitation of the FETi and SETi (see Sects. 2.8.2 and 4.2.7.1). However, stretching the chordotonal organ during active movements of a restrained animal also briefly inhibits these neurons (see Sect. 3.2.2). Thus, the response of animals in which the thorax has been almost completely denervated corresponds to that of animals which are struggling to extricate themselves from a restraining harness but not to that of walking animals.

Comparison of these results with those discussed in Sect. 4.2.6.3 leads to the conclusion that the femoral chordotonal organ is by itself unable to drive the motor output of a leg.
5. In further experiments the neck connectives were also cut in a denervated thorax preparation. The motor output of protractor and retractor motor neurons after abdominal stimulation is the same with or without intact neck connectives. Regular sequences do not occur less frequently. Since removal of the subesophageal ganglion abolishes walking in otherwise intact animals (see Sect. 4.2.6.4) this is another indication that the motor output recorded from the denervated thoracic ventral nerve cord does not correspond to "walking" (Bässler and Wegner, in prep.).

In summary the motor output of animals with a denervated thoracic CNS corresponds not to walking but rather to irregular searching or extricating movements. Thus, the organization of the program governing movements of a single leg during walking cannot be purely central. Certain peripheral influences seem to comprise an indispensable part of the program. At least in the protractor-retractor system there appears to be a central mechanism which is capable of producing some rhythmic oscillation. Such a mechanism could be termed a central oscillator (see Sect. 2.5.4). The oscillator for each leg and probably the oscillators for muscle groups within a leg are not coupled. These central oscillators are likely to be part of the walking program, but this has not yet been proven.

Another possible alternative is that there is a central oscillator for the whole program "walking" of a leg, which is capable of producing a rhythmic motor output "walking" by itself, but which does not function under the given circumstances. This suggestion is not very probable, because another central oscillator (the rocking oscillator) functions under the circumstances (Sect. 2.5.5).

These findings appear at first to be contradictory to those of Pearson and Iles (1970) who always registered a regular rhythmic output from the completely deafferented ventral nerve cord of the cockroach. Delcomyn and Daley (1979), however, report that also in cockroaches a regular rhythm is seen only in a small fraction of the preparations. They do not describe what happens in the larger fraction. This difference between cockroaches and stick insects may simply be due to the fact that cockroaches have a high walking frequency so that there is not enough time available for peripheral influences (Graham 1979b).

4.2.6.2 Unspecific Stimulation of Sense Organs

If sensory input is an integral component of the program for the walking movement of a leg, particular sense organs must be able to drive the motor output all by themselves. An observation from an earlier experiment suggested a possible basis for testing this supposition. When the retractor coxae muscle of a middle or hindleg is either denervated or cut at its insertion (keeping innervation of the muscle and all other nerves intact), the operated leg moves normally during treadwheel walking (Graham 1977a). However, in this situation the backward movement of the leg during retraction is passive. The leg is dragged back by the motion of the treadwheel which is being driven by the other legs. The motor output in nl_2 (protractor) and nl_5 (retractor) is relatively normal (Graham, in prep.).

In the actual experimental set-up a motor-driven conveyor belt was mounted next to the treadwheel, level with the hindleg which was the only leg studied. The belt was driven at a very low speed. The hindleg that is moved slowly backwards by the belt executes far fewer steps than the other legs. As the leg is being pulled backwards, the nl_5 (retractor neurons)

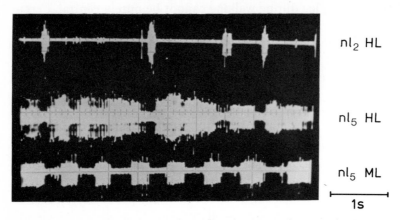

Fig. 4.21. Recordings from nl_2 (protractor) and nl_s (retractor) of a right hindleg and from nl_s of a right middle leg. The retractor coxae muscle of the right hindleg has been cut at its coxal insertion, and the hindleg is walking on a conveyor belt while the other legs are walking on a treadwheel. The hindleg step frequency is distinctly lower than that of the other legs (compare nl_s recordings for middle and hindleg). The nl_s activity of the hindleg is modulated in the step frequency of the middle leg

shows high activity which is modulated in the step rhythm of the other legs. A strong protractor burst brings the leg forwards (Fig. 4.21). It is usually set down again on the conveyor belt so that the whole process starts over from the beginning. A completely intact leg shows similar behavior under the same conditions (unpublished).

4.2.6.3 Stimulation of Single Receptor Organs

The conveyor belt experiment demonstrates that sense organs are apparently able to influence the timing of the motor output. The effect of an individual sense organ can only be determined if a defined stimulus is used. However, presentation of a defined stimulus is a difficult proposition. Electrical stimulation of a complicated mechanoreceptor with various types of sense cells is an unnatural and ill-defined stimulus. Defined mechanical stimulation of most sense organs is only possible in an immobilized leg. Even when the leg is held rigidly in one position, muscle contractions provide undefined stimuli to the campaniform sensilla and to any tension receptors which may be present. For this reason the influence of the motor output on muscle and cuticle tension must be neutralized before a sense organ is stimulated. Since all of the larger leg nerves are mixed, this can be done only by operating on the muscle. Two surgical procedures were followed.

1. According to Section 4.2.4.2 *campaniform sensilla* situated on the trochanter and the proximal end of the femur seem to be able to drive the motor output. Before mechanical stimulation of these receptors an operation was performed on a middle leg in which the protractor and retractor coxae muscles were denervated and the tendons of all leg muscles were cut at their insertions. Nerves nl_3 and nl_4, the autotomy muscle, and the nervus cruris in the femur were also severed in this leg. The operated leg dangled loosely while the other legs walked on a treadwheel. The campaniform sensilla were stimulated by a rhythmic pressure on the trochanter and the femur base of the operated leg. In some animals such stimulation can drive the motor output of this leg's nl_2 (protractor neurons) and nl_5 (retractor neurons) in a rhythm which is independent of the walking rhythm of the other legs. Pressure on the trochanter is accompanied by nl_5 activity, release of pressure by nl_2 activity (unpublished).

In another preparation the entire thoracic ventral nerve cord was denervated (see Sect. 4.2.6.1) with the exception of the stimulated middle leg, which was operated on as described in the preceding paragraph. Recordings were made from the stumps of nl_2 and nl_5 of this leg as well as from nl_2 of the ipsilateral hindleg. As in the case of the totally denervated thoracic ventral nerve cord (see Sect. 4.2.6.1), no coordination is seen after abdominal stimulation. However, rhythmic stimulation of the campaniform

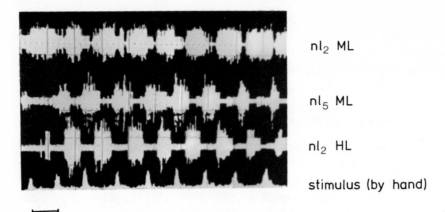

nl₂ ML

nl₅ ML

nl₂ HL

stimulus (by hand)

1s

Fig. 4.22. Recordings from nl_2 (protractor) and nl_5 (retractor) of a middle leg and the nl_2 of the ipsilateral hindleg. The thoracic ventral nerve cord was entirely denervated except for the middle leg which was operated on as described in the text. The campaniform sensilla on the trochanter were stimulated rhythmically by hand *(bottom trace)*

sensilla drives not only nl_2 (protractor neurons) and nl_5 (retractor neurons) activity but also nl_2 activity in the ipsilateral hindleg (Bässler and Wegner, in prep.; see Fig. 4.22). Stimulation of the sensilla excites retractor neurons and prevents strong activity of the protractor neurons. This agrees with the results from experiments using a metal clip to continuously stimulate the campaniform sensilla of intact animals (see Sect. 4.2.4.2).

The motor output is now more similar to that of walking animals. Activity in neighboring legs is coordinated. Since the common inhibitor could not be positively identified in these recordings, no conclusion can be drawn as to its action. There seems to be only one essential difference with respect to walking animals. The protractor burst is not divided into two parts. This conforms with the hypothesis that attributes the protractor burst at the beginning of a stance phase in intact legs to a resistance reflex.
2. The influence of *position receptors on the subcoxal joint* was studied by amputation of a hind- or middle leg just prior to the distal end of the coxa. This operation removes the insertion of the coxa muscles as well as the campaniform sensilla. While the animal was on a treadwheel, the position receptors were stimulated by moving the coxa stump backwards and forwards with a motor.

In the hindleg of a standing animal there is normally a negative feedback for both directions of movement (see Sect. 3.4). When the animal is walking, this negative feedback persists during the passive forwards movement (Fig. 4.23, brief nl_5 activity during the forward swing). During the backwards movement there is also strong nl_5 (retractor neurons) activity, which serves to reinforce the externally applied backward movement,

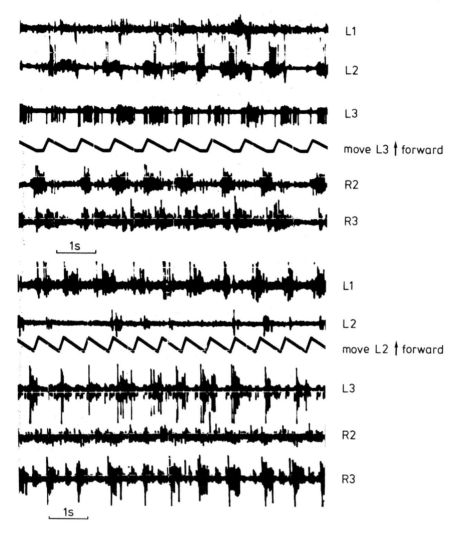

L1

L2

L3

move L3 ↑ forward

R2

R3

1s

L1

L2

move L2 ↑ forward

L3

R2

R3

1s

Fig. 4.23. Myograms for the retractor coxae muscles of several legs. In the *upper picture* the fourth trace shows the movement of the stump of leg L3. The animal is walking on a wheel. In the *lower picture* the stump of leg L2 is moved *(third trace)*. (Graham)

functioning as a positive feedback. What was a negative feedback in the resting animal, thus, becomes a positive feedback in the walking animal. This is a further example of a *program-dependent reaction* (see Sect. 3.2.3).

The middle leg also has a positive feedback for backwards movements. In this leg, however, the motor output is far more strongly modulated by the walking movements of the other legs (Fig. 4.23; Graham, unpubl.).

4.2.6.4 Elimination of Higher Centers

The experiments discussed in this section are aimed at determining to what extent higher centers participate in the control of walking.

Decerebration (removal of the supraesophageal ganglion) or cutting of the esophageal connectives increases walking activity without substantially altering walking movements (for details, see Sects. 2.2 and 4.2.1). After decapitation or cutting of the neck connectives the animals are no longer able to walk. Cutting the neck connectives or the other thoracic connectives on only one side does not prevent the animal from walking. If both connectives between the pro- and meso- or the meso- and metathoracic ganglia are cut, the legs posterior to the cut usually no longer participate in walking (v. Buddenbrock 1921). Thus, at least a unilateral connection to the subesophageal ganglion seems to be a prerequisite for walking.

In addition to a central release of walking, there is also a peripheral release as shown by the following experiments.

1. Even when all nervous connections between meso- and metathoracic ganglia are cut, intense abdominal stimulation can occasionally elicit stepping movements by the hindlegs. The hindlegs are pulled backwards relative to the body by the walking movements of the other legs, lift up, and swing forward, etc. However, if a piece of smooth paper is placed under the hindleg tarsi, the legs drag this paper along the table and no longer lift up (unpublished).
2. When the connectives between the pro- and mesothoracic ganglia are cut, the forelegs walk normally as long as the other legs are pulled along and do not resist this movement. However, if the mechanical load on the forelegs is changed by cutting off the body at the level of the cut connectives, coordinated walking is no longer observed (unpublished).
3. When the left connective between pro- and mesothoracic ganglia is cut and the right middle leg and left hindleg are amputated, the left middle leg is often just dragged along on smooth surfaces where it cannot get a good grip. On a rough substrate, it moves with the other legs although with a smaller amplitude (v. Buddenbrock 1921).

4.2.6.5 Conclusions

A thoracic ganglion produces a rhythmic output corresponding to walking only if it is connected to the subesophageal ganglion. The subesophageal ganglion does not appear to directly control the rhythm. Rather it seems to have only a nonspecific stimulatory effect. This is supported by three findings. First after denervation of the thoracic nerve cord no rhythmic output is produced which corresponds to walking. Activity corresponding to rocking can still be produced (see Sect. 2.5.5). Secondly, the motor output can be driven by a rhythmic sensory input. Finally, the influence of the subesophageal ganglion can be replaced by an especially strong abdominal stimulation.

The central nervous portion of the control network for a leg's walking movement cannot produce a motor output corresponding to walking without sensory inputs either from that leg or the other legs. These findings can be interpreted in two ways. Either there is a centrally organized walking program which is only activated by the presence of unspecific afferences. Or peripheral components are an integral part of the walking program. It is not yet possible to make a final decision between these two alternatives, but so far all findings speak more for the second possibility. There is no regular rhythm in restrained animals (constant sensory input). On the other hand afferences from a leg cannot only brake its walking rhythm (especially delay the transition to the succeeding phase, see Sect. 4.2.4) but also speed it up or drive it (see Sect. 4.2.6.3). Whether the control program for a step is a simple reflex chain or a more complex program is discussed later (see Sect. 4.2.7.1).

When one leg is stationary or moves more slowly than the others, an oscillation in the step frequency of the other legs is superimposed on its motor output (see especially Sect. 4.2.5). In a denervated ganglion, this is reflected in a rhythmic oscillation of the motor output (Sect. 4.2.6.1). Since it has been shown that the origin of this oscillation must lie outside of the ganglion half in question, it is tempting to regard it as an expression of the coordinating influences from other legs.

4.2.7   Responses to Irregularities of the Walking Surface

Every phase of a step must be adjusted to surface irregularities for an animal to move about effectively in its natural environment. Walking differs in this respect from other modes of locomotion like flying or swimming which do not require such a high degree of accommodation. This is probably one of the chief reasons why technology has long been able to successfully mimic flying and swimming. Walking machines in comparison are available only in very imperfect models even though walking is superior to every technical form of terrestrial locomotion for traversing rough ground.

The experiments which have been discussed so far have ignored the problem of environmental irregularities. This problem will be examined in this chapter.

4.2.7.1   Control of Femur-Tibia Joint Position During Walking

The feedback control loop of the femur-tibia joint has been thoroughly investigated in inactive animals (see Sect. 2.4.2, 2.5.3, 2.6 and 2.8). During active movements by restrained animals, it is switched off and replaced by a positive feedback for flexion (see Sect. 3.2).

Is this feedback control loop operating in a walking animal? To clarify this question experiments were performed in which animals were allowed to walk freely on a bar, a small section of which could be moved by hand (see Fig. 4.24). When a leg (middle or hindleg) stepped on this platform, it was either pushed in or pulled out. This stimulus was presented at various times during a stance phase but only once per step. At the same time the electrical activity of nerve F2 containing the extensor motor neurons (for techniques, see Sect. 2.8.2) and myograms from the flexor tibiae

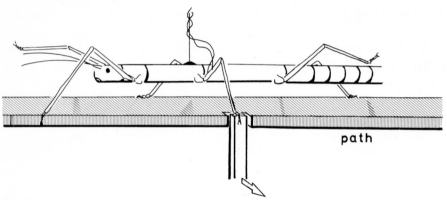

Fig. 4.24. Experimental set-up for examining the femur-tibia feedback control loop in a walking animal. When a leg steps on the platform, the platform is pushed in or out. The activity of nerve F2 and the flexor tibiae muscle are recorded. (Cruse)

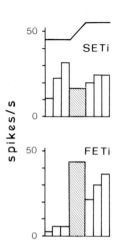

Fig. 4.25. The *SETi* and *FETi* activity during pushing in of the platform in the set-up shown in Fig. 4.24. The width of the white columns equals 100 ms. (Cruse and Pflüger 1981)

muscle were recorded. The spike frequencies of the FETi and SETi were used for quantitative analysis. Figure 4.25 presents the averaged results of numerous recordings during pushing in of the platform. The results show that the stimulus elicits a response in the form of a negative feedback, i.e., the control loop is in operation during walking. The FETi responds strongly during flexion of the joint. In a resting animal it is the SETi that primarily responds. The fact that the SETi frequency does not appear to change could be a masking effect. A small SETi spike is often not identifiable near a large FETi spike. The response decays faster in a walking animal than the step response in a resting animal (Cruse and Pflüger 1981). This is not necessarily peculiar to the walking animal. In a resting animal the response to a step stimulus superimposed on sinusoidal stimulation of the femoral chordotonal organ (which approximates the walking situation) also declines relatively quickly (Bässler 1972b).

Myograms of the flexor show that pulling out the platform elicits responses mainly from the units with large impulses (Cruse and Pflüger 1981). In this case small impulses may also be masked.

Similar experiments were performed on animals on a treadwheel where the conditions could be better defined (Schmitz 1980). A hindleg stood off to the side of the wheel on a movable platform. The set-up was comparable to that shown in Fig. 4.14. Pulling out and pushing in the platform for a standing animal produces the expected resistance reflexes. These reflexes are also present in the walking animal. Measurement of the force developed by the leg reveals that the control loop gain is on the average smaller by a factor of three in a walking animal than in a standing one. The electrophysiological results show that the change in gain applies to all units as long as their impulse frequency is not near the upper saturation boundary as is often the case for the SETi in a walking animal.

In all units the response decay is significantly faster during walking than during standing. Since in these experiments, the femur-tibia joint was only passively moved, the faster response decay in walking animals must be due to an alteration of the system in the walking state (program-dependent reaction).

The results demonstrate that the femur-tibia control system is operating during walking. This system apparently functions very differently according to varying behavioral situations. In a resting animal the gain can be very variable. It increases after a disturbance and is higher directly after an active movement which has been triggered by a disturbance (Fig. 2.12). During walking and directly after spontaneous active movements by a restrained animal the gain is low (Fig. 2.12). During active movements of restrained animals the control loop is switched off (gain is not measurable) and replaced by a positive feedback for flexion. As already mentioned in Sect. 3.2.3 this is another example of a *program-dependent reaction*.

During walking the response to stretching the chordotonal organ is different than it is during active movements by a restrained animal. This finding provides an insight into the organization of the walking program itself. It requires that if the program is simply a chain of reflexes, the activity of particular reflex pathways (if the animal can walk) must determine the direction and gain of other reflex pathways, e.g., from the femoral chordotonal organ to the extensor motor neurons. This far exceeds what is commonly meant by a reflex chain. It more benefits the organization of the walking program to regard it as a type of computer program that requires sensory feedback at certain points to continue running.

4.2.7.2 Load Increase

When an animal is walking uphill or gets caught on something, the load on its legs is greater than when it is walking on a horizontal surface. Walking under an increased load produces increased muscle forces (see Sect. 4.2.2) and an increased forward extension of the legs, i.e., the anterior extreme position is shifted forwards (see Fig. 4.4). This response is at least partially controlled by sense organs of the same leg. This conclusion is based on the observation that the force exerted on a stationary force transducer by the leg of an animal walking on a wheel (experimental situation Fig. 4.14) is considerably higher than the force which a freely moving animal exerts on the ground (see Sect. 4.2.2).

The campaniform sensilla are at least partially responsible for this behavior because the intensity of the retractor output increases according to the pressure on the trochanter (see Sect. 4.2.6.3). Whether position receptors also play a role in this response is still unresolved. This would be the case if each joint were controlled by servomechanisms and is compatible with the way the femur-tibia control loop functions during walking. If position receptors do participate, a deviation from the reference position resulting from increased loading should lead to an increase in the motor output. This has not been directly tested. However, two experimental results speak against it. When middle legs with crossed receptor apodemes (see Sect. 4.2.4.1 and 4.2.5.2) execute a step, they move their half of a double wheel more slowly than do intact legs. This implies that at least in this situation a servomechanism is of subordinate importance (Graham and Bässler 1981). Secondly, passive backwards movements of the subcoxal joint increase the motor output to the retractor coxae muscle. This corresponds to a positive feedback, not to the negative feedback required by a servomechanism (see Sect. 4.2.6.3).

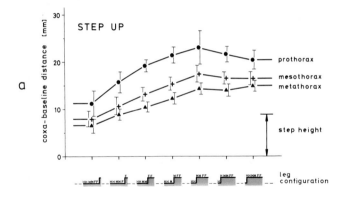

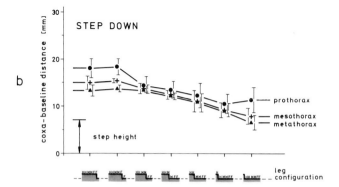

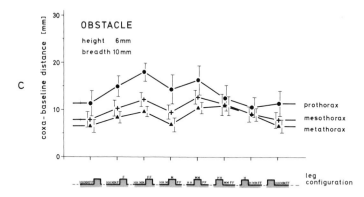

Fig. 4.26a–c. The heights of the coxae of the three thoracic segments when the insect is walking over a **a** "step up," **b** "step down," and **c** "obstacle." The values are for the different leg configurations as shown on the abscissa. *Vertical bars* give the standard deviations of the corresponding mean values. (Cruse 1976b)

4.2.7.3   Obstacles During the Swing Phase

When the femur of a middle or hindleg runs into an obstacle while swinging forward, the leg is not simply set down. Instead it reverses its direction of movement and then swings forward again. The whole sequence is repeated with a frequency of 3 Hz—4 Hz until the obstacle is surmounted (Cruse 1980a).

This behavior is probably related to "saluting" by legs with crossed receptor apodemes (see Sects. 4.2.4 and 4.2.5.2), which can be similarly interpreted. The chordotonal organ in legs with crossed receptor apodemes incorrectly reports that the femur-tibia joint is flexed although the extensor muscle is strongly activated. Under normal conditions this would mean that something is obstructing extension. Lifting the leg at the coxatrochanter joint is a "meaningful" response for getting around such an obstacle.

4.2.7.4   Control of Segment Height (Cruse 1976b)

Free-walking stick insects were filmed as they walked over various kinds of irregularities (step up, step down, obstacle, ditch). The height of the coxa of each segment was measured (Fig. 4.26). Except for very high upward steps which are practically a transition between horizontal and vertical walking surfaces, the meso-metathoracic joint hardly moves when an animal walks over an obstacle. The body can thus be regarded as approximately rigid under these conditions. Normal body posture with the prothorax highest and the metathorax lowest is maintained even on the obstacle. In walking over an uneven surface the longitudinal body axis forms approximately the same angle with the average substrate slope in the region of the body as it forms with the horizontal plane during walking on a level surface.

The results of these experiments can be represented by the same model used to describe height regulation by standing animals. In this case it is also unnecessary to assume neuronal channels from one segment to the other. However, just because a model without intersegmental neuronal channels can describe the results within the framework of measurement accuracy does not exclude the possibility that such channels exist in a walking animal (see Sect. 4.3.5).

4.2.8   Model for the Control of a Single Leg

Cruse (1980a) has constructed a model that sufficiently describes most of the results presented in Section 4.2. It was designed for the control of a generalized insect leg during walking and, therefore, does not take into account some peculiarities of the stick insect. Also, it was conceived as a building block or subsystem for a model to describe the coordination of the movements of all six legs.

The model for the control of walking movements of a single leg contains a central rhythm generator, which is constructed like a relaxation oscillator. It can produce a rhythmic output after elimination of the whole periphery, a feature which is probably not applicable to the stick insect. As long as the periphery remains intact, sensory commands decide when the leg enters the next phase of its step cycle. The switching from central to peripheral control of leg movement is achieved in the model by the fact that the afferences indicate the deviation from a reference value instead of the actual leg position. Thus, the absence of a sense organ results in the signal "no deviation from reference value," and the model behaves as if the reference value were reached. It is an interesting special case of a central-peripheral program in so far as elimination of the periphery scarcely changes the motor output frequency. It also raises a note of caution that even if deafferentation does not alter the motor output in a real system, the program cannot be assumed to be purely central.

The relaxation oscillator, whether or not it is controlled by the periphery, produces the actual reference value for the control system that determines leg position. It is still open whether such a servomechanism exists for stick insects (see Sect. 4.2.7.2). Coordinating influences from other legs change the threshold of the oscillator, as does the periphery, and influence its output.

Since a few basic assumptions of the model either do not normally apply to the stick insect or have not been definitely demonstrated (e.g., organization of the central part or the assumption that sensory systems measure deviation from a "desired" position), its heuristic value for investigating the neural basis of the control mechanisms for a single leg is limited. However, hypotheses for the control mechanism of a single leg are necessary for tackling the problem of leg coordination control. Since this model satisfactorily describes a large part of the experimental data, in the absence of a better model it is useful as a building block for constructing a complete model to describe the coordination of all six legs.

4.3   Interactions Between Legs that Do Not Influence Timing

4.3.1   Hindleg "Aims" at Tarsus of Middle Leg

Free-walking animals set the hindleg tarsus down right behind and slightly to the outside of the tarsus of the ipsilateral middle leg. This response occurs irrespective of where the middle leg tarsus is. Thus, the position of the middle leg determines where the ipsilateral hindleg steps.

An experiment by Cruse (1979b) illustrates this response particularly well. An animal was fastened over a treadwheel as shown in Fig. 4.14. The right middle leg stood on a rigid horizontal platform (larger than in Fig. 4.14.). Touching the animal's abdomen caused it to walk. When the right hindleg stepped on the platform after its first swing phase, the distance between the tarsi of the middle and hindleg was measured. In these experiments the tarsus of the middle leg stood in five different positions relative to the animal's body. No matter where the middle leg tarsus is, the hindleg tarsus steps on the average right behind and slightly to the outside of it (Fig. 4.27).

Several other findings corroborate this result. In one experiment stick insects are forced to walk over a 15 mm wide ditch, which is so deep that the animals cannot reach the bottom with their legs. When the legs step into the ditch, they make searching movements until they have found an edge, usually the front edge of the ditch. Forelegs step with 65%, middle legs with 48%, and hindlegs with only 25% probability into the ditch. The probabilities with which the fore-, middle, and hindlegs step on a 15 mm wide band which has been painted on a horizontal surface are 60%, 56%, and 56% (Cruse 1979b). Obviously the hindlegs step into the ditch much less often than they step onto a bar of the same width. This experiment suggests an adaptive purpose for the response. Where the middle leg has found a solid foothold, the hindleg will with a high probability also find one.

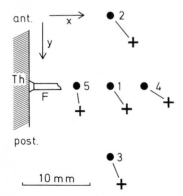

Fig. 4.27. The five test positions of the middle leg tarsus *(dots)* relative to the insect's thorax *(Th)* and the mean points for each of these test positions where the hindleg tarsus reached the platform *(crosses)*, viewed from above. The proximal part of the femur of the middle leg is shown schematically *(F)*. (Cruse 1979b)

When a middle leg with crossed receptor apodeme of a treadwheel walking animal executes a normal stance phase, the ipsilateral hindleg often tries to set down off of the wheel (Graham and Bässler 1981). The femur-tibia joint of the middle leg is quite flexed in the stance phase. Therefore its femoral chordotonal organ reports that the joint is extended. Apparently information from the femoral chordotonal organ of the middle leg is used to guide the placement of the hindleg.

When the hair plates and rows on the subcoxal joint of the middle leg are manipulated so that they continuously signal a protracted leg position (see Sect. 4.2.4.3), the hindleg is set down in front of the middle leg tarsus (Fig. 4.13). Thus, the sensory hairs of the middle leg also appear to influence the placement of the hindleg tarsus.

4.3.2 Treading-on-Tarsus (TOT) Reflex (Graham 1979b)

The placement of the hindleg tarsus right behind and to the side of the ipsilateral middle leg tarsus has a certain variability. For this reason it can happen, although rarely, that the hindleg treads on the middle leg tarsus. This phenomenon was investigated in slowly walking decerebrated animals where it occurs more frequently. If a hindleg steps on the ipsilateral middle leg tarsus, the hindleg is immediately raised and set down again just behind the middle leg. If the middle leg tarsus is touched lightly with a fine brush directly after the hindleg has been correctly set down, the hindleg is again raised briefly and set down a little farther back. This behavior is called the treading-on-tarsus or TOT reflex. The TOT reflex has a latency of 105 ± 20 ms. It is also seen occasionally in intact animals at stepping frequencies under 2 Hz.

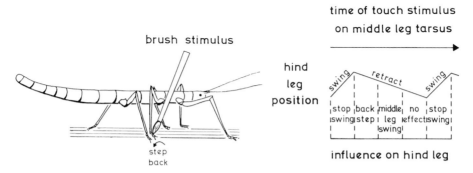

Fig. 4.28. The treading-on-tarsus (TOT) reflex. *Left* the experimental set-up; *right* the response to middle leg stimulation during different phases of the hindleg step. (Graham)

The TOT reflex can only be elicited directly after the hindleg has been set down (Fig. 4.28). If the hindleg is in a later part of its stance phase, no response occurs. When the hindleg is in the middle part of its stance phase, the middle leg is in its swing phase and cannot be stimulated. If the brush stimulus is applied during hindleg swing phase, the leg is set down prematurely.

Thus, the hindleg response depends on which phase of a step cycle the leg is in. This is a typical phase-dependent reaction which is a special case of the *program-dependent reaction* (see Sect. 3.2.3).

4.3.3   Increased Loading

Increasing the load an animal has to overcome increases the intensity of the motor output and shifts the anterior extreme position forwards (see Sects. 4.2.1.1 and 4.2.7.2). Not only a leg's own sense organs (Sect. 4.2.7.2) but also those of the other legs play a role in the response. This has been shown in experiments on first instar larvae of *Extatosoma* in which the animals walked on a wheel with one leg on a force transducer (see Sect. 4.2.5.1; Bässler 1979). When the load on the walking legs is increased by braking the wheel, the leg on the force transducer is often lifted and set down again very far forwards on the wheel. If the leg remains in place on the force transducer, the force exerted on the force transducer increases sharply. This behavior is the same no matter which leg is on the force transducer. Similar results have also been reported for *Carausius* adults (Cruse and Saxler 1980a).

4.3.4   Influence of Searching Movements on the Other Legs

When a middle leg with crossed receptor apodeme of a treadwheel walking animal executes a stance phase, the ipsilateral hindleg frequently makes searching movements off of the wheel (see Sects. 4.2.5.2 and 4.3.1). During these searching movements, the movement of the other legs either ceases or slows down (Graham and Bässler 1981).

A "salute" by a leg with crossed receptor apodeme can be thought of as a prolonged swing phase with superimposed searching movements (see Sect. 4.2.5.2). If the middle leg starts to salute, the next protraction of the ipsilateral foreleg does not take place 30% of the time (Fig. 4.16; Graham and Bässler 1981). Furthermore, the average anterior extreme position of the foreleg is shifted posteriorly in free walking animals with operated middle legs (Bässler 1977b). It appears as if the probability of other legs being lifted is lowered if one leg is in the air (see also Sect. 3.5). Since part of the experiments were carried out on a treadwheel, it is unlikely that this effect is due to the increased loading on the other legs. This mechanism may participate in the control of coordination.

4.3.5 Control of Body Height During Walking (Bässler 1977b)

Do the contralateral legs influence each other when an animal is walking over a step with only one side of its body? This question was investigated using animals attached dorsal side down to a small carriage which could be moved on horizontal rails (Fig. 4.29). Above them were two sticks which were parallel to the rails and one of which was interrupted to form a step. The force exerted on each of the two sticks was recorded as the animal walked along them.

Decerebrated animals were tested first. When an animal walks over a step with only one side of its body, there is a compensation on both sides of the body. The forces are directed as if the animal were "trying" to bring all the tarsi to the same level.

When the hair plates BF1 (see Sect. 6.2.2) are removed from the step side of the body, the compensation for this side is almost as great as before the operation but the compensation for the other side disappears. Cutting the receptor apodemes in the legs on the step side decreases the response of the contralateral side but does not eliminate it completely. These results show that the contralateral legs are influenced by afference mainly from the hair plates BF1 and perhaps weakly from the femoral chordotonal organs.

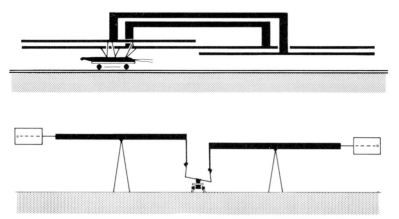

Fig. 4.29. Experimental set-up for testing the control of body height. *Top drawing* shows a side view without scale beams; *bottom* an anterior view with scale beams and transducers

Intact animals walk very quickly in this set-up if at all. A compensation can only be observed on the step side. From these experiments it is not possible to decide whether the compensation on the opposite side is a result of decerebration or slow walking.

Compensation on the non-step side is not seen in standing animals (see Sect. 3.6). In any case, whether or not this response is associated with slow walking or decerebration, it is dependent on which program is running at the time. It is, therefore, a further example of a *program-dependent reaction.*

4.3.6 Amputation of Single Legs

Amputation of single legs changes primarily the coordination of the remaining legs (see Sect. 4.4.2), but it can also somewhat alter the way these legs are moved. For instance, amputation of a foreleg shifts the anterior extreme position of the ipsilateral middle leg rearwards. Amputation of a middle leg shifts the posterior extreme position of the ipsilateral foreleg quite far back. These are the only significant effects that single leg amputation has on the extreme positions of the other legs (for details of recording method and significance criteria see Bässler 1972a).

When the hair plates and rows on the subcoxal joint of either a fore- or middle leg are manipulated so that they continuously signal that the leg is swung far forward (see Sect. 4.2.4.3), amputation of the leg directly behind it increases the percentage of stance phases which are not completed. At the end of a stance phase the tarsus must be jerked from the substrate by the other legs (see Sect. 4.2.4.3). This demonstrates that amputation of a hindleg also impedes the transition from stance to swing phase in a middle leg (Bässler 1977b). This hindleg effect was not detected in the earlier amputation experiments apparently because of the high significance criteria which were used.

The results of these experiments are difficult to interpret because amputation changes so many parameters, e.g., sense organs are removed, the stump moves only slightly and with a different coordination relative to the other legs, the coordination of the other legs changes, the remaining legs carry an altered load. Thus, the results only provide evidence that effects do occur, both forwards and backwards, that can alter the extreme positions.

4.4 Coordination

4.4.1 Description of Leg Coordination–Gaits

4.4.1.1 Nymph Walking

Free walking first instar nymphs use two distinctly different gaits (Graham 1972). Figure 4.30 illustrates the more frequently used tripod gait (Gait I). Each swing phase is represented by a black bar. The time scale reads from left to right. Such a record facilitates the quantification of the following features which are used to characterize a gait.

Period = duration of a step (stance plus swing phase measured from swing phase onset).

Lag = time interval between the beginning of a swing phase in one leg and the beginning of a swing phase in another. The lag between R3 and L3 is written $_{R3}L_{L3}$.

Phase = the lag between two legs divided by the period of the first. This makes the value for lag independent of period duration. The phase of L2 relative to R2 is written L2:R2.

Figure 4.30 shows that in the tripod gait the ipsilateral fore- and hind-legs and the contralateral middle leg (e.g., L1, L3, and R2, see swing phases surrounded by dotted line in Fig. 4.30) are swung forward simultaneously. It is typical of this gait that both legs of one segment have a relative phase of 0.5, which means that they are exactly alternating. Also, swing

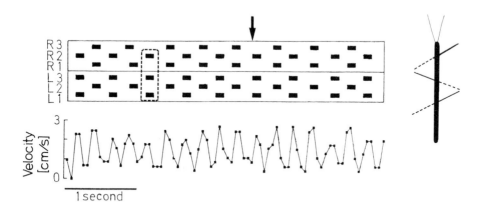

Fig. 4.30. The step pattern of a first instar stick insect walking with a tripod gait (Gait I). Legs on the right *(R)* and left *(L)* are numbered, *1, 2,* and *3* from front to rear. *Black bars* indicate when the leg is in a swing phase. *Dotted lines* enclose legs that are swung forward simultaneously. Picture on the *right* schematically shows the position of the insect at the time indicated by the *arrow* (legs in the swing phase are *dashed*). *Bottom trace* gives the velocity of the animal during the walk. (Graham 1972)

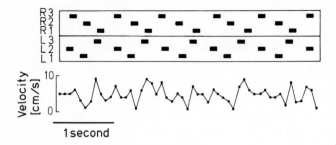

Fig. 4.31. The step pattern of a first instar stick insect walking with a tetrapod gait (Gait II). For explanation see legend of Fig. 4.30. (Graham 1972)

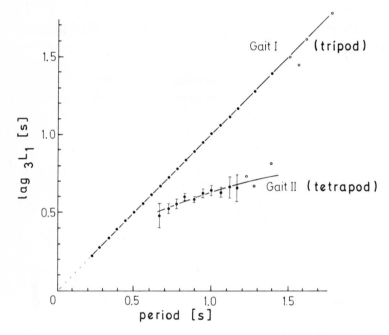

Fig. 4.32. Metachronal lag ($3L_1$) versus period for 10 first instar stick insects. *Closed circles* represent the mean value of $3L_1$ for a given mean period. *Error bars* denote the error of the mean where this is larger than the symbol. (Graham 1972)

phase duration is proportional to the period (see Sect. 4.2.1.1). In the tripod gait coordination is always exact and is maintained even when the animal is changing direction. Turns are produced by changes in step length rather than step frequency.

The second gait of first instar larvae is illustrated in Fig. 4.31. Since no more than two legs are ever protracting at the same time, at least four legs are always on the ground. For this reason it is termed the tetrapod gait (Gait II). In contrast to the tripod gait, swing phase duration for the tetrapod gait is independent of the period. Also, the coupling between left and

right body sides is very labile, and the coordination appears somewhat irregular. The animal rarely maintains a straight course for very long in the
tetrapod gait. Turning is achieved primarily by increasing the step frequency of the legs on the outside of the turn and decreasing it on the inside. While the tripod gait is used throughout the animal's whole range of
walking speeds, the tetrapod gait is limited to slow walking. The two gaits
are clearly distinguishable in a plot of the lag between ipsilateral hind- and
forelegs versus period as shown in Fig. 4.32.

Third instar nymphs also use primarily the tripod gait (Jander and
Wendler 1978). Turning in this instar is carried out usually by a change in
the step length with maintenance of coordination.

4.4.1.2 *Free Walking Adults* (Graham 1972)

At slower walking speeds the gait of the adult corresponds to the
tetrapod gait of the first instar (Fig. 4.31). Swing phase duration is also independent of the period. The adult walk is much more regular than the
tetrapod gait of the first instar. Long straight or slightly curved walks are
frequent. The lag between ipsilateral hind- and forelegs is given as a function of period in Fig. 4.33 for both body sides. Comparison of Figs. 4.32

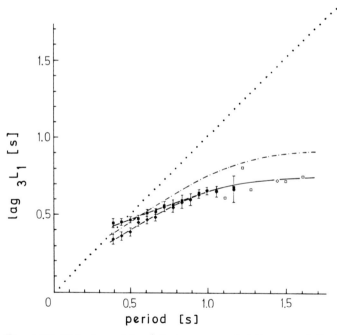

Fig. 4.33. Metachronal lag $(3L_1)$ versus period for one adult stick insect. *Squares* represent left leg steps; *diamonds* right leg steps. *Dashed-dotted line* is derived from the
results of Wendler (1964) for an adult insect walking on a treadwheel. *Dotted line*
represents Gait I for the first instar stick insect. (Graham 1972)

and 4.33 makes it clear that the adult gait is similar to the tetrapod gait of the first instar nymph. These similarities justify classifying the adult gait as tetrapod.

At the highest adult walking speeds, the metachronal sequences (back, middle, front) overlap. Hindleg protraction occurs simultaneously with or even slightly before foreleg protraction instead of after as during slow walking. Since the transitions are gradual, it is questionable whether the adult uses a distinct tripod gait.

During high speed walks the left and right legs of the same segment have a phase of 0.5, i.e., they are exactly in antiphase. This contrasts with slow walks where right leg protraction follows protraction of the left leg in the same segment. The phase R:L is about 0.35. The temporal asymmetry is almost always in this direction; the reverse asymmetry is rare. The temporal asymmetry of the system is also reflected in the different curves for right and left legs shown in Fig. 4.33.

During a turn the asymmetry may shift in the other direction. The coupling between left and right sides can weaken or disappear completely. This decoupling may explain why a temporal asymmetry in the tetrapod gait of the first instars, which are always changing direction, could not be demonstrated. Also it may facilitate the increase of step frequency on the outside of the turn.

4.4.1.3 Adult Walking on a Treadwheel

Wendler (1964, 1978) used a treadwheel composed of two rigidly connected wheels. The distance between animal and treadwheel was fixed. Femur position for all six legs was recorded automatically. Using this apparatus Wendler (1964) made the first quantitative analysis of leg coordination in the stick insect. His results are in essential agreement with those discussed in Section 4.4.1.2, with one exception. He finds in general no temporal asymmetry except possibly at low step rates. Figure 4.33 shows the dependence of lags $_{R3}L_{R1}$ and $_{L3}L_{L1}$ on the period. The somewhat higher values may be due to the altered mechanical loading of the legs.

Graham (1981a) measured leg coordination on a treadwheel composed of two independent wheels with very low friction. The left legs walked on one wheel and the right legs on the other (see Sect. 4.2.1.2). The leg coordination on such a wheel corresponds to the tripod gait over a wide range of walking speeds (Fig. 4.34). The animal can spontaneously decouple the two body sides and walk with different frequencies on the left and right sides. This probably corresponds to turning.

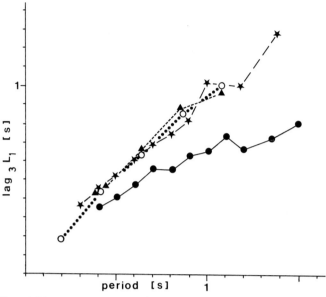

Fig. 4.34. Metachronal lag *(3L₁)* versus period for different experimental situations. *Triangles* represent walking on a double treadwheel with very low friction (Graham 1981); *stars* the same experiment by Foth (unpubl.); *closed circles* another experiment by Foth using the same animals on a wheel with increased friction; *open circles* walking on a mercury surface. (Graham and Cruse 1981)

4.4.1.4 Walking on a Mercury Surface

Legs which are simultaneously on the ground or on a treadwheel are mechanically coupled. To eliminate the effects of this substrate coupling, adults were made to walk on a mercury surface (see Sect. 4.2.1.3; Graham and Cruse 1981). The leg movements of these animals are only about 70% coordinated (compared to about 90% for animals on a rigid substrate). The well-coordinated walks closely resemble the tripod gait of first instar nymphs (Fig. 4.34).

4.4.1.5 Dependence of Coordination on the Resistance

These experiments employed a double treadwheel as described in Graham (1981a) which was modified so that the friction of each wheel could be raised independently (Foth, unpubl.). Animals walking on such a wheel use the tripod gait with low wheel friction and the tetrapod gait with high wheel friction (Fig. 4.34). Swing phase duration depends on period duration at low friction but not at high friction (see Sect. 4.2.1.2). When the friction on the left and right wheels is different, one body side may walk with a tripod gait and the other with a tetrapod gait. Since usually the right and left step frequencies also differ when this occurs, it often leads to decoupling of the two sides. Double steps on one side and gliding coordination between the body sides may also be observed. These findings show that external influences such as the resistance which has to be overcome determine which gait will be used for walking.

4.4.2 Coordination After Surgical Intervention

In all the preceding chapters a description of a particular behavior was followed by a presentation of experiments that were meant to provide insight into the mechanisms controlling these behaviors (see Sect. 1.3). Such experiments are particularly powerful when a single defined input unambiguously produces a single quantifiable output. As yet such experiments are not possible for the *investigation of control mechanisms of leg movement coordination.* It has been known since v. Buddenbrock (1921) that each leg possesses its own control mechanism and that these individual control mechanisms interact with each other. Thus, the problem of leg coordination can be phrased in terms of how these six control mechanisms influence each other and over what channels.

There are two strategies for attacking this problem: (1) Surgical operations can be used to disrupt the system in as many ways as possible. The information obtained from such experiments can be used to construct a model which behaves as much as possible like the real system. This model can then be used to make predictions which can be tested in new experimental situations or at least to suggest experiments aimed at verifying assumptions implicit in the model. Like all attempts to solve a problem with methods of systems theory (see Sect. 2.9), this strategy has primarily a heuristic character. Another drawback is that it is often not obvious how a particular surgical operation on the animal should be represented in the model. (2) The system is given a single defined input, e.g., a defined stimulus to a sense organ. At present the effect of such inputs on the motor output and the control mechanism of the same leg is only approximately known if at all. Clear statements about coordination mechanisms cannot be made until such effects are known. Before 1979 there was no known way to impose a rhythm on the motor output of one leg which was independent of the movement of the other legs (the data reported in Sects. 4.2.6.2 and 4.2.6.3 are all new). For this reason all earlier studies were carried out according to the first strategy.

In the experiments discussed in the remainder of this section the animals were physically altered by amputation of single legs or by other surgical operations. The effect of these operations on coordination was then recorded.

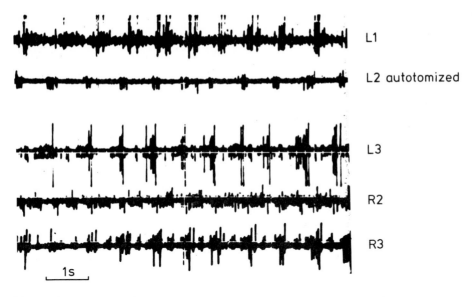

Fig. 4.35. Myograms of the retractor coxae muscles of different legs during walking on a treadwheel. The left middle leg has been autotomized so that only the coxa and trochanter of this leg are present. (Graham)

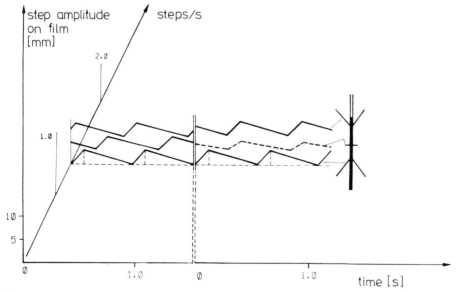

Fig. 4.36. Step pattern and leg coordination as a function of the step frequency. The phase relationships between ipsilateral legs before *(left)* and after *(right)* autotomization of both middle legs. The movement of the middle leg stump with decreased amplitude is indicated by *heavy dashed line*. (Wendler 1978)

Leg Amputation. It has long been known that amputation or autotomization of particular legs qualitatively alters the coordination of the remaining legs (v. Buddenbrock 1921). Wendler (1964) improved these experiments in two respects. He made quantitative measurements, and he used animals firmly supported over a treadwheel so that the loading of the remaining legs was not changed by amputation of one leg.

The clearest effect of amputation is seen after middle leg amputation. As shown in Fig. 4.36 the foreleg has now the same phase relative to the ipsilateral hindleg as the middle leg had in the intact animal. The stump of the middle leg continues to receive a weak rhythmic output (Fig. 4.35), which produces small amplitude movements. The observed coordination between hind- and forelegs after middle leg autotomization does not exist anywhere in the whole spectrum of walking speeds in the intact animal. The coupling between fore- and hindlegs seems to be rather weak and the legs can walk with different frequencies. When this occurs, gliding coordination can be seen (Fig. 4.37).

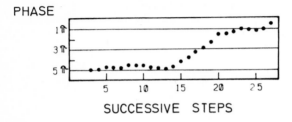

Fig. 4.37. Phase relationship between right fore- and hindlegs of a stick insect which has autotomized both middle legs. *Even π* legs are in phase; *odd π* legs alternate. At the beginning the legs alternate. From the 15th step on the front leg adopts a slightly higher stepping frequency, so that the phase relationship changes progressively (gliding). Fixed alternation recurs from the 20th to the 26th step

Attachment of a balsa wood "leg" to the middle leg stump restores the "intact" coordination (Wendler 1966). Thus, the altered coordination after middle leg autotomization is not caused by elimination of afferents on the femur, tibia, or tarsus.

Graham (1977a) performed a large number of amputations on first instar nymphs. His results may be summarized as follows: All five- or four-legged animals select a gait that is the same as or very similar to the tetrapod gait, but with a pronounced right-left coupling. This is not surprising since the tripod gait would lead to unstable configurations in which the animal is standing on only two legs. Of the two possible asymmetric forms of tetrapod gait, the animal chooses either the form in which it does not ever have to stand on only two legs or it alternates between the two forms. Exceptions to these rules occur when both asymmetric forms lead to situations where the animal must stand on two legs, e.g., after amputation of both middle legs. All the one and two leg amputations (only contralateral legs) lead to stable coordination except for amputation of both middle legs where gliding coordination between fore- and hindlegs occasionally occurs.

Leg Immobilization. Graham (1977a) also immobilized single legs by gluing them to either the head, thorax, or abdomen of first instar larvae. The results of these experiments differ significantly from those in which the same legs were amputated. If the middle leg is only immobilized, the fore- and hindlegs on this side are not significantly coordinated.

Crossing of the Receptor Apodeme of a Right Middle Leg (Graham and Bässler 1981). Figure 4.16 shows that as long as the right middle leg with crossed receptor apodeme (see Sects. 4.2.4.1 and 4.2.5.2) is walking normally, the coordination resembles that of the intact animal. The contralateral phase relationships are expressed by $R:L > 0.5$. When the leg starts to "salute", the first protraction of the ipsilateral foreleg is often omitted. During the remaining part of the salute the coordination resembles that of a middle leg amputee.

Legs on Force Transducers (Cruse and Saxler 1980a, b). When one or several (maximally five) legs are standing on separate force transducers and the others are walking on a treadwheel, one can study either the coordination of the walking legs, as Wendler (1964) did, or the coordination of the force oscillations (see Sect. 4.2.5.1) with each other and with the walking legs. Cruse and Saxler (1980a, b) made the plausible assumption that the force maximum corresponds to a stance and the minimum to a swing phase. Figure 4.38 shows the results in the form of phase histograms relative to a walking reference leg. The absence of a distinct peak in a histogram means that this leg is not coordinated with the reference leg. It may, however, be coordinated with other legs. Two aspects of the results are particularly noteworthy. First, the coordination between walking and standing legs only rarely corresponds to the coordination of intact animals or amputees. Secondly, the force oscillations of adjacent legs standing on force transducers are often in phase when walking legs are in antiphase.

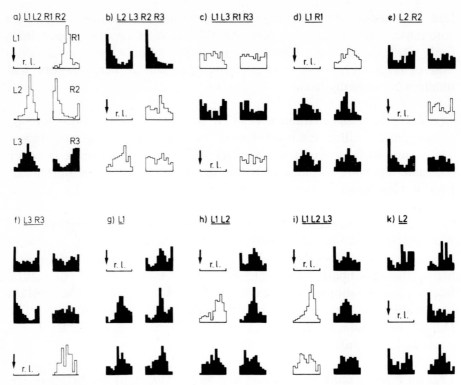

Fig. 4.38a–k. The coordination between legs walking on a treadwheel *(white)* and the force oscillations of legs standing on separate force transducers *(black)*. The phase histogram of the investigated leg relative to a walking reference leg *(r.l.)* is shown for each of the experimental situations (**a–k**). The reference point for a walking leg is the onset of retraction (indicated by *arrow* for the reference leg); and for a standing leg, it is the force maximum. (Cruse and Saxler 1980b)

4.4.3 Coordination Models

Four models have been proposed up to now: (1) Wendler (1968a, 1978), (2) Graham (1972, 1977b), (3) Cruse (1979a) and (4) Cruse (1980a, b). The first three are relatively simple compared to the last one. All of the models are able to describe coordination in intact animals at changing walking speeds. None of them describes all the experimental data.

All of the models use a relaxation oscillator to represent the control mechanism of one leg. In model (1) one leg influences another via an additional input to the integrator of the oscillator. In models (2) and (3) one leg alters the threshold of the oscillator of another leg.

Wendler's model is composed of six relaxation oscillators. They are coupled in such a way that contralateral legs of the same segment influence each other with the same intensity. Ipsilateral legs influence each other from back to front. The hindleg influences the middle leg strongly and the foreleg weakly. Control of walking speed occurs via a non-rhythmic input common to all the oscillators. It is one of the integrator inputs and influences integration velocity.

In addition to intact walking coordination the model can also describe the coordination of animals with amputated middle legs. Asymmetric coordinations are not represented; they occur in the experimental set-up used by Wendler only at low walking speeds (see Sect. 4.4.1.3).

Graham's model uses a normal relaxation oscillator with a variable threshold to simulate the control mechanism of a single leg. When the threshold is reached, a protraction of the leg begins and the oscillator is reset to zero, beginning a new cycle. The higher the input, the higher the frequency of this oscillator and thus, the step frequency.

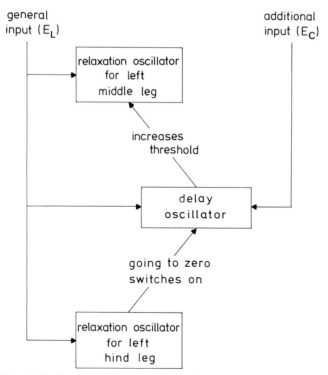

Fig. 4.39. Two relaxation oscillators (each responsible for the control of a single leg) are coupled over a delay oscillator. For details see text

Two such relaxation oscillators are coupled so that resetting the first raises the threshold of the second. This prolongs the stance phase of the second leg (Fig. 4.39). To obtain the experimentally determined phase relationships of adjacent legs, this increase in threshold must have a certain duration. This is achieved in the model by inserting a delay oscillator between them which executes only one cycle each time its threshold is reached. It is turned on when the first relaxation oscillator is reset to zero. As long as the delay oscillator is running, the threshold of the following relaxation oscillator is raised. The delay oscillator threshold can be adjusted to match the experimentally obtained values for the delay of retraction onset in the second leg. The input for all three oscillators is the same and determines the walking speed. A system with six relaxation oscillators coupled in this manner shows all the features of the tripod gait as long as the frequency of each oscillator is higher than that of the one immediately behind it (Fig. 4.40). The coupling of the relaxation oscillators for one body side must be from rear to front with the hind- and middle legs mutually influencing each other contralaterally.

An additional low level input to the delay oscillators which is constant for all walking speeds (E_C in Fig. 4.40) causes the model to switch to the tetrapod gait. When inputs to the oscillators on the left and right sides differ (E_L and E_R in Fig. 4.40), the system reproduces the typical temporal across-the-body asymmetry found in free-walking insects.

This model can describe both walking gaits, both forms of asymmetrical coordination, and changes in walking speed, gait, and direction. Many of the amputation results can be replicated by adding a weak influence of the hindleg on the ipsilateral foreleg. All of the amputation experiments can be simulated by assuming an additional mechanism that decides which

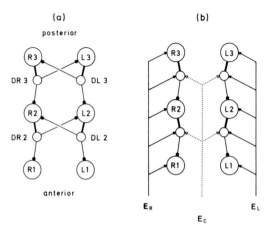

Fig. 4.40a, b. The organization of the complete model. a The interaction between the leg oscillators proposed for the intact stick insect. b The leg and delay oscillators and their respective inputs E_R, E_L, and E_C. (Graham 1977b)

of the possible gaits or asymmetries is the optimal solution for each combination of remaining legs. This model is also capable of simulating coordination in other insects, even those whose hindlegs sometimes walk with different frequencies (Graham 1978). The only experimental results it cannot reproduce are the coordination of force oscillations of standing legs when others are walking (Cruse and Saxler 1980a, b).

Cruse's "simple" model functions completely differently from the two preceding ones. Wendler and Graham, as have many other authors (Hughes 1957; Wilson 1966), started with the observation that there is a metachronal wave of swing phases from rear to front on each side (Fig. 4.41, solid line). This led to the assumption that information flow is directed anteriorly. A closer look at such a step sequence, reveals a second metachronal wave in the opposite direction (Fig. 4.41, dotted line). Cruse's "simple" model incorporates this observation and is constructed with an information flow from front to rear.

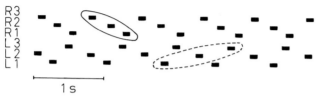

Fig. 4.41. Typical coordination pattern of an adult stick insect. (Graham 1972)

In this model two relaxation oscillators are coupled in such a way that when a certain below-threshold value (switch position) is reached in one oscillator, the threshold for a second oscillator is lowered. This results in the second oscillator reaching its threshold earlier. In neurophysiological terms this kind of influence can be termed excitatory. This contrasts with the inhibitory influence in the two preceding models. The setting of the switch position depends on the walking speed. The direction of the influence is shown in Fig. 4.42. Influences running transversely and longitudinally (t_1) favor leg alternation while diagonal influences (t_2 favor leg synchronization.

This model describes coordination in both gaits. It always produces the same stable coordination no matter in which leg position it starts. However, it is unable to simulate the results of amputation experiments or the coordination of force oscillations in legs standing on force transducers. This model was extended to account for these findings.

Cruse's extended model is based on the simpler version. The control mechanism of one leg is represented by the model described in Section 4.2.8. It

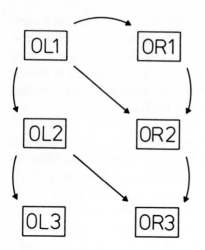

Fig. 4.42. Cruse's "simple" model. *OL1* oscillator for the left foreleg; *OR1* for the right foreleg; *OL2* for the left middle leg; etc. *Arrows* indicate the direction of excitatory influences. (Cruse 1979a)

includes the influence channels t_1 (excitatory influence with antiphase bias) and t_2 (excitatory influence with in-phase bias) plus two further channels, t_3 and t_4. The former channel is an inhibitory influence with antiphase bias, similar in effect and arrangement to that of the models by Wendler and Graham. Channel t_4 is an excitatory channel with in-phase bias, which raises the force in the controlled leg when the load of the controlling leg is very high (Fig. 4.43). Since t_4-channels only come into play when the leg is heavily loaded, these channels are normally turned off in the walking leg. They are able to simulate the altered and often in-phase coordination of force oscillations between legs on force transducers. The protractor activities shown in Fig. 4.22 may be a direct demonstration of these channels.

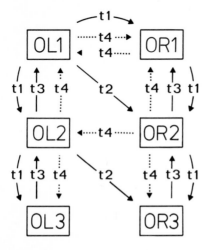

Fig. 4.43. Cruse's "extended" model (1980b). For details see text

This model can simulate the coordination of free walking animals in only one of the two asymmetric forms. For the mirror image asymmetry the t_1 channel between the forelegs and all the t_2 channels must run in the opposite direction. Most of the results from experiments with legs on force transducers can be reproduced by the model. Its applicability for amputation experiments has not been tested yet because it is not clear how to represent amputation in the model.

Critical Evaluation of the Models. All of the above models assume that leg coordination is produced exclusively by the interaction of six control sub-systems, each of which is responsible for one leg. Neurophysiologically this means that coordination is produced exclusively in the thoracic ventral nerve cord. Only Graham (1977b) takes into account the possibility that higher centers may decide which form of asymmetry is optimal for a given amputation configuration and then act on the system according to the decision. Observations on decerebrate walking suggest that such an intervention by higher centers is possible. The walk of decerebrated adult stick insects is well-coordinated. Although the coordination is rigid for a particular walk, it can differ for different walks and different animals. The values for the lags $_{R3}L_{R1}$ and $_{L3}L_{L1}$ (see Sect. 4.4.1.1), as well as for swing phase duration (see Sect. 4.2.1.1) lie somewhere between those for the normal tripod and tetrapod gaits. Decerebrates can walk in either of the two asymmetrical forms of the tetrapod gait. These findings could be explained if the supraesophageal ganglion has an influence on which gait is used, i.e., determines which of the available coordinating channels is to be used in each individual case.

Other findings also suggest that the channels used for coordination may vary according to the situation. After amputation of one middle leg the forelegs sometimes walk almost synchronously, i.e., in the same phase (Graham 1977a). On the other hand in the force transducer experiments the forelegs are in antiphase when they are walking and the other legs stand still (see Fig. 4.38d). The forelegs also walk in antiphase if the connectives between the pro- and mesothoracic ganglia are cut. These last two experiments require coordinating channels with antiphase bias between the forelegs. However, the middle leg amputation experiment suggests that there are no coordinating channels between the forelegs. In this case the forelegs are coordinated by influences from the middle and hindlegs.

The reported findings indicate that certain coordinating channels may be switched on or off according to the situation. If this hypothesis can be substantiated by further experiments, it would have two important consequences: (1) No model can simulate all experimental findings without a mechanism to switch single channels on or off according to the situation.

In Cruse's extended model the t_4 connections already have a kind of selection mechanism which only turns them on at higher loads. (2) The control of coordination is considerably more complex than has been assumed.

4.5 Walking Backwards (Epstein and Graham, in prep.)

Pinching the antenna of an animal on a treadwheel or a mercury surface often causes it to walk backwards. Such walks can be relatively long-lasting and regular. In contrast to normal forwards walking the protractor coxae and depressor trochanteris muscles are simultaneously active as are the retractor coxae and levator trochanteris muscles. The different coordination of muscles, however, is not the only change. Many "reflexes" also reverse their direction. During normal walking hampering the retraction by placing a rod on the middle of the femur increases retractor muscle activity (see Sect. 4.2.7.2, increased leg loading). During backwards walking the same hindrance leads to immediate interruption of swing phase followed by several back and forth swings. Obstructing leg protraction with a vertical rod held at mid-femur height immediately interrupts protraction in a swing phase of forward walking and induces rapid back and forth swinging (see Sect. 4.2.7.3). During a stance phase in backwards walking, which is now a protraction, the same stimulus produces an increase in protractor muscle activity. This is a further example of a *program-dependent reaction*.

5 Orientation

5.1 Gravity Orientation

5.1.1 Description of Behavior

As soon as a blinded *Carausius* adult encounters a vertical rod, it climbs to its tip. If the upper end of the rod bifurcates, the stick insect climbs to the tip of one of the branches where it gropes around in the air with its forelegs, then turns around and goes up the other branch. This behavior is repeated several times before the animal eventually climbs back down the rod (Precht 1942). Apparently the stick insect has a negative geotaxis which temporarily switches over to a positive geotaxis after the end of a branch is reached. The more tips encountered, the longer the duration of the positive geotaxis. Such behavior may serve to bring the insect into a favorable feeding site.

When a stick insect is given cones of varying steepness to climb, it walks towards the tip, then turns around and walks downwards. The steeper the cone, the less the insect's walking directions deviate from the vertical (Schneider 1961).

5.1.2 Receptors Involved in Behavior

All terrestrial insects that have been investigated measure the angles between their body axes and the direction of gravity by proprioceptors which register the load on individual joints due to the effect of gravity. These proprioceptors which function as gravity receptors can under some circumstances replace each other so that it is difficult to identify them using an either—or response test. The method introduced by Yagi (1928) of turning tendency comparison was selected for quantifying the directional effect of gravity.

Adult stick insects *(Carausius)* on a vertical walking surface were illuminated from one side with the light parallel to the surface. The insects walked at least one body length before their subsequent resting positions were recorded. Figure 5.1 illustrates the recorded angles: α is the angle

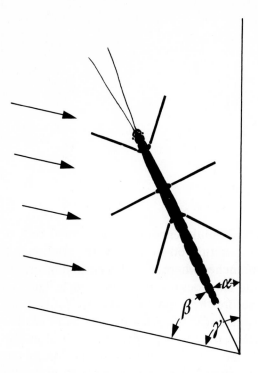

Fig. 5.1. Illustration of the angles relevant to this chapter. *Arrows on the left* indicate the incident direction of light

between the longitudinal body axis and gravity; β, the angle between longitudinal body axis and incident light direction; and γ, the angle between direction of incident light and gravity ($\gamma = \alpha + \beta$).

The resting direction assumed by most of the insects is a compromise between the direction of light and gravity (Fig. 5.2, α_1). The few individuals that come to rest turned away from the light (α_2) or with their heads

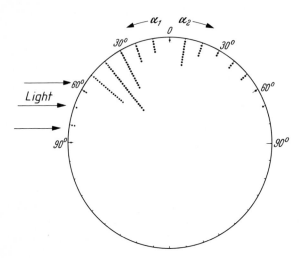

Fig. 5.2. Resting positions assumed by *Carausius* on a vertical surface under 60 W illumination from the side ($\gamma = 90°$). 100 measurements from five animals

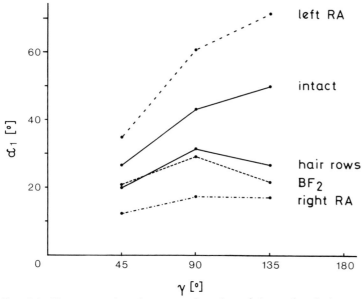

Fig. 5.3. The mean of angle α_1 as a function of the angle γ in intact animals and after ablation of various sense organs. Illumination is from the left with constant intensity. *RA* cut receptor apodeme in all legs of one body side. Coxal hair rows and hair plates *(BF_2)* were removed from all six legs

pointing straight down, were excluded from further observations (for detailed justification see Bässler 1965).

The mean of the angle α_1 increases with light intensity and is, of course, also a function of γ (Fig. 5.3). It is unaffected by immobilization of all intersegmental joints or removal of the antennae. However, when either the hairs of BF2 or the hair rows on the coxae of all legs are shaved off, the mean of α_1 decreases. Cutting the femoral receptor apodemes of legs on the shadow side of the body also decreases α_1, whereas the same operation on the illuminated side increases α_1 (Fig. 5.3, Bässler 1965). When the receptor apodemes on the illuminated side are attached to the cuticle so that the chordotonal organs are maximally and continuously stretched, α_1 decreases (Bässler 1967).

These results demonstrate that the sensory hairs on the subcoxal joint and the femoral chordotonal organs participate in the perception of the direction of gravity. This effect may be indirect since ablation of these receptors may lower the muscle tone of individual muscles leading to decreased stimulation of campaniform sensilla (Wendler 1972).

Wendler (1965b) showed that sense organs on the antennae also participate in gravity perception. He fastened stick insects to a holder which was attached to a counterweight hung over a pulley. The animals were free to turn around their vertical axes (Fig. 5.4). When the holder weight is

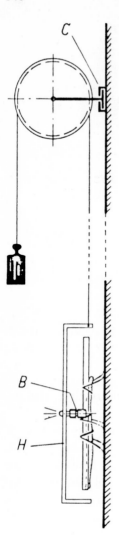

Fig. 5.4. Experimental set-up to reverse the direction of the force exerted on the legs by the body. The weight (0.7 g) of the balsa wood holder *(H)* is always counterbalanced by a counterweight. A miniature ball bearing at *B* allows the animal to turn freely around its vertical axis. *C* indicates the rails along which the pulley moves. (Wendler 1965b)

compensated by the counterweight, the animals walk upwards in the dark on a vertical surface just like free walking animals. If the counterweight is increased so that it pulls the insect upwards with a force corresponding to twice its body weight, some animals walk upwards and some downwards. However, when in addition the antennae are amputated, all the insects walk downwards.

5.1.3 Processing of Proprioceptive Input

Two directionally oriented physical parameters, gravity and light, act on an animal on a vertical surface. These can serve as references for setting the walking direction or body position. Under experimental conditions most animals attempt to orient their longitudinal axis not only in the direction of gravity (negative geotaxis) but also towards the light source (positive phototaxis). If these two directions are at an angle to each other, the insect assumes a resultant of the two. It is in equilibrium when the turning tendency due to negative geotaxis (D_S) exactly counterbalances the turning tendency due to positive phototaxis (D_L), i.e., when

$$D_S = -D_L$$

How do the signals from the proprioceptors on the subcoxal joints, femur-tibia joints, and antennae combine to form D_S?

The antennal proprioceptors appear to affect only the sign but not the magnitude of D_S for the following reasons. When light impinges from the side, amputation of the antennae does not alter α_1. If, in addition, an upwards pull is exerted on the center of gravity of these animals, they all walk downwards. Their deviation from vertical due to positive phototaxis remains the same as in unloaded, intact animals under the same lighting conditions (Bässler 1971).

The mean of α_1 is very reproducible for intact stick insects walking on a vertical surface with side illumination. Gluing additional weight to the insect's center of gravity does not measurably change the mean of α_1. In fact, the standard deviation of the mean decreases as body weight increases. The mean of α_1 on an inclined walking surface is the same as on a vertical one, but the standard deviation is markedly increased. In animals with no sensory hairs on the subcoxal joints or with cut receptor apodemes on one body side, α_1 is a function of body weight (Fig. 5.5). Apparently the weight of the body is taken into account by each individual component and these components are then processed in such a way that the body weight is cancelled out. The simplest mathematical operation for such a process would be the formation of a fraction. The sense hairs on the subcoxal joints measure the force in the direction of the longitudinal body axis. Under the experimental conditions this is G (body weight) · cos α (Fig. 5.6). Since the legs are on the average held vertically to the body, the sense organs on the femur-tibia joints approximately measure the forces in the direction of the transverse axis. This is G · sin α (Fig. 5.6).

Removal of the sense hairs on the subcoxal joints decreases G cos α. Since D_S increases, G · cos α must be in the denominator (for details of the derivation of the model, see Bässler 1965). This gives:

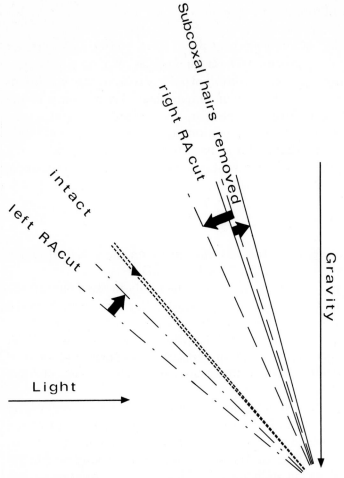

Fig. 5.5. The effect of increased load on the mean of α_1 for animals on a vertical walking surface before and after removal of sense organs. Two means connected by an arrow are given for each operation. *Base of the arrow* lies on the value from unloaded insects, and *arrow tip* points to the value for animals, whose body weight was increased by one-third

$$D_S = K \frac{G \sin \alpha}{G \cos \alpha} = K \tan \alpha$$

where K is a constant. This equation is consistent with all the experimental findings:

1. After the hairs on the subcoxal joints are shaved off, $G \cdot \cos \alpha$ is replaced by a small constant. D_S is larger and increases with subsequent increases in G.

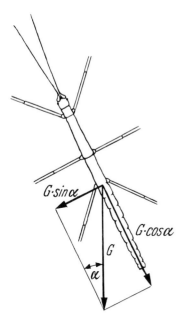

Fig. 5.6. Distribution of forces on a vertical surface for animals that are not aligned with the direction of gravity. G body weight

2. After the receptor apodemes on the shadow side are cut, half of component $G \cdot \sin \alpha$ is replaced by a large constant (after the operation the chordotonal organs report maximally extended femur-tibia positions). Thus, D_S increases (α_1 becomes smaller). An increase in G causes D_S to become smaller since G is weighted more heavily in the denominator than in the numerator.

3. Cutting the receptor apodemes on the illuminated body side or continuous stretching of the chordotonal organs on the shadow side have exactly the opposite effect (Bässler 1965, 1967).

4. When the angle γ is changed for the intact insect,

$$\frac{\sin \beta}{\tan \alpha_1} = \text{constant}$$

How are the sensory signals from the antennae incorporated in the calculation? Figures 5.7 to 5.9 show the resting positions of insects lacking either antennae or sense hairs on the subcoxal joints or both (Bässler 1971). The antennal sense organs affect the sign but not the magnitude of D_S. Apparently the antennae and the sense hairs on the subcoxal joints can substitute for each other with respect to the sign of D_S. If only one type of sense organ is removed, the animals can still distinguish between up and down. They only lose this ability when both receptor types are missing. This can be represented in the model by multiplying $G \cdot \sin \alpha / G \cdot \cos \alpha$ by +1 when the proprioceptors on the subcoxal joints and

Fig. 5.7. Resting positions of 15 stick insects after amputation of the antennae. $\gamma = 90°$ from the left

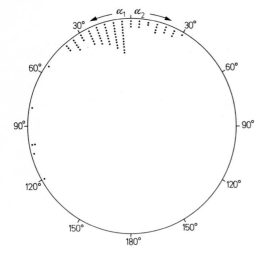

Fig. 5.8. Resting positions of 15 stick insects after removal of the sense hairs on the subcoxal joints. $\gamma = 90°$ from the left

antennae report "head up" and by -1 when they report "head down." When both proprioceptor types are missing or when their signals are contradictory and cancel each other out (e.g., upwards pull with doubled body weight),

$$D_S = K \frac{\sin \alpha}{\cos \alpha} = K \tan \alpha$$

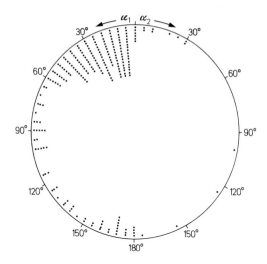

Fig. 5.9. Resting positions of 10 stick insects after removal of antennae and the sense hairs on the subcoxal joint. $\gamma = 90°$ from the left

The function $y = \tan x$ transects the X axis at $0°$ and $180°$ in the same direction (i.e., y' has the same sign for $x = 0°$ and $x = 180°$). The insect would then have two stable positions, one at $0°$ and the other at $180°$. However, since in the intact insect the head down and head up positions have opposite signs, the function crosses the abscissa at $0°$ and at $180°$ in opposite directions. Thus, only $0°$ is a stable position; $180°$ is a labile position (for details and a discussion of menotaxis see Bässler 1975).

The switchover from negative to positive geotaxis that is observed when the insect reaches the tip of a branch could be accomplished by changing the signs of the components produced by the antennae and the subcoxal joints. This also explains why the standard deviations for cone ascent and descent are the same (Schneider 1961) since the ratio $G \cdot \sin \alpha / G \cdot \cos \alpha$ is formed in the same way for both cases. The standard deviation decreases with increasing body weight because the absolute error in the measurement of the components remains constant and its relative contribution decreases as G increases. (The standard deviation is attributable to measurement errors and not to menotaxis. Bässler 1965).

5.2 Light Orientation

5.2.1 The *Carausius* Eye

Carausius does not have any ocelli. The compound eyes are small and are typical apposition eyes (Jörschke 1914). In the first instar the eye has 250–300 ommatidia. The angle of divergence between two ommatidia in the horizontal plane is about $7.5°$ (Kalmus 1937). The adult eye has 300–

400 ommatidia (Jander and Volk-Heinrichs 1970), and the divergence angle between ommatidia has been reported to be 3°–5° (Bauers 1953) and 5°–7° (Jander and Volk-Heinrichs 1970).

The electroretinogram is typically monophasic. ERG recordings reveal that the eye has a wide adaptation range. The dark-adapted eye is about 22,000 times more sensitive that the light-adapted eye. This corresponds well to the adaptation range of the human eye. Recovery time is 30 min. In the dark-adapted state the *Carausius* eye is at least 700 times as sensitive as that of *Calliphora*. Depending on the experimental conditions, the fusion frequency is between 7 and 40 Hz, being lower at lower light intensities (Autrum 1950). (For the sensitivity spectrum of the eye and extraocular light perception see Sect. 2.2).

5.2.2 Photomenotaxis

Carausius exhibits a distinct photomenotaxis (see also Sect. 5.1). The menotactic angle can be practically any size and is easily changed by external influences such as a movement by the experimenter. If one light is turned off and another one on, an angle of 12° between the two lights is sufficient to elicit a change in orientation. The angle of this compensatory turn is often greater than the angular distance between the two lights (v. Buddenbrock 1931).

5.2.3 Optomotor Response – Color Vision

The optomotor response of free walking first instar nymphs was used to test their ability to perceive colors. An insect is placed in a drum with alternating stripes of yellow and gray. Drums with different shades of gray are used to determine which gray could no longer be distinguished from yellow. This gray is then combined with different shades of blue until the animal no longer responds to the rotating drum. The animals are then presented with a drum composed of this blue and the original yellow. Insects with good color vision such as bees still show an optomotor response. Since first instar nymphs do not respond to this or to seven other color combinations, it was concluded that they do not possess color vision (Schlegtendal 1934). Experiments on body color change (e.g., Priebasch 1933) have led to the same conclusion.

5.2.4 Orientation to Visual Patterns

When first instar nymphs are placed in a circular arena with striped vertical walls, they walk towards the black-white boundaries. They are attracted by narrow black stripes on a white background as well as narrow white stripes on a dark background. After reaching the arena wall they follow the painted stripes upwards, even if the stripes are slanted relative to the vertical. This behavior is thought to guide stick insects from the ground up into trees or bushes (Kalmus 1937).

Jander and Volk-Heinrichs (1970) used an alternate choice test for the relative attractivity of various black-white patterns for adult stick insects. They also found a preference for vertical striping. The details of their results are difficult to interpret.

5.3 Idiothetic Orientation

Idiothetic orientation enables an animal to maintain a course even in the absence of direction giving external cues such as gravity and light (Mittelstaedt 1978). Burger (1972) demonstrated that stick insects can orient idiothetically. Animals were placed in the middle of a horizontal arena under red light. After they had walked 15 cm, either a light source was turned on or the arena was tilted by 90°. This orienting external cue elicited a course change, referred to here as the light or gravity deflection. After the animals had walked about 10 cm in their new course, the external stimulus was removed. The animals responded by making counter-

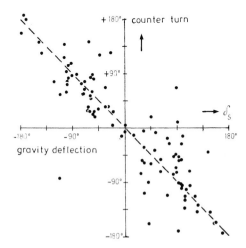

Fig. 5.10. Counterturns following a gravity-induced course change. After the stick insect walked 10 cm on the horizontal arena under red light, the arena was tilted 90° around a fixed axis. After the insect had walked a further 10 cm, the arena was returned to its original position. 97 measurements from 10 animals. (Burger 1972)

turns which were approximately proportional to the light or gravity deflection (Fig. 5.10). The insects are apparently able to register the degree of turning elicited by the external cue and to compensate for it when the external cue is removed. Since in all experiments on orientation to external stimuli (e.g., Sect. 5.1) the insects had executed some turns before they were tested, idiothetic turning tendencies were probably present. This is one, if certainly not the only, source of the great variance in the data from orientation experiments.

6 Anatomy of the Muscles, Nerves, and Sense Organs of the *Carausius* Thorax

The general anatomy of *Carausius* has been described by Bauchhenss (1971). Further details on the muscles and nervous system may be found in Marquardt (1940). This chapter covers only the anatomy that is necessary for the understanding of the preceding text.

6.1 Anatomy of the Thorax and the Legs

6.1.1 Skeletal Components and Joints

The three thoracic segments differ in size. The prothorax is relatively short and is a typical wingless thoracic segment. The meso- and metathorax differ from those of other insects in their lack of wings, the reduction of the parts of the cuticle on which the wings are normally found, and their great elongation. Since the elongation is anterior to the attachment of the legs, the coxal region is shifted to the posterior end of each of these segments. The first abdominal segment is fused to the metathorax and thus belongs functionally to the thorax.

The joint between the pro- and mesothorax is very flexible and in normal behavior is moved both horizontally and vertically. The joint between meso- and metathorax can be bent passively to an angle of about 40°. Normally, however, the insect keeps it almost rigid even when walking over large obstacles. Angles of up to 30° can be actively produced only under extreme conditions such as in the transition from horizontal to vertical walking and in searching movements when the metathorax is fixed (see Sects. 3.6, 4.2.7.4; Cruse 1976b).

All the legs are relatively long and flexible and are similarly constructed. The coxa has a normal dorsal articulation with the thorax at the subcoxal joint. Its ventral articulation is formed by the trochantin which is movable relative to the body. The mobility of this strut allows the ventral pivot of the coxa to be moved towards (proximally) or out from (distally) the body thus changing the position of the rotational axis of the subcoxal joint. During walking the rotational axis varies by 20°–50°(Cruse 1976a) (Figs. 6.1, 6.2).

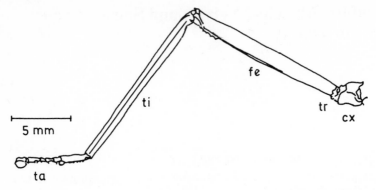

Fig. 6.1. Anterior view of the right hindleg of *Carausius*. The leg is comprised of the coxa *(cx)*, trochanter *(tr)*, femur *(fe)*, tibia *(ti)*, and tarsus *(ta)*. (Modified from Bauchhenss 1971)

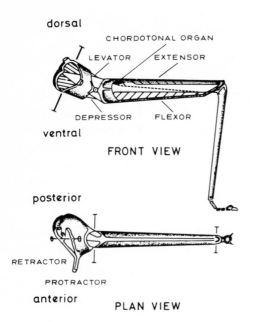

Fig. 6.2. Schematic drawing showing the articulation and some of the musculature that controls joint movement in the left leg of a stick insect. The tendons of the protractor and retractor coxae muscles are only partially shown. (According to Graham)

The coxa and trochanter form a hinge joint with its rotational axis almost exactly parallel to the longitudinal body axis when the leg is perpendicular to the body.

The trochanter and femur are fused. The leg can easily autotomize at their boundary where a membrane is formed which prevents the loss of body fluids.

Indentations at the base of the prothoracic femurs fit around the head when the legs are extended forwards. The femur-tibia joint is also a hinge joint with its rotational axis parallel to that of the coxa-trochanter joint.

The tibia-tarsus joint and the joints within the tarsus itself are movable in all directions (for details on joint mobility see Cruse 1976a).

6.1.2 Musculature

The subcoxal joint is moved primarily by the protractor and retractor coxae muscles. The fibers of both muscles originate from the side wall of the body anterior to the coxa and are relatively long (up to 6 mm in the retractor coxae muscle). The *retractor coxae muscle* consists of three separate fiber bundles with a common tendon which attaches to the posterior rim of the coxa. Bundles Ra and Rb insert at the tergum, 5–6 mm anterior to the rotational axis of the joint. In the mesothorax the third bundle, Rc, has its insertion near this axis so that when the middle leg is in the posterior extreme position, Rc can exert a force perpendicular to the tendon. In the metathorax Rc inserts a few millimeters anterior to the rotational axis so that all three bundles work in the same direction in all hindleg positions (Graham and Wendler 1981a).

The protractor coxae muscle lies ventral to the retractor. Its tendon attaches to the soft cuticle immediately in front of the anterior rim of the coxa.

Due to the reduced size of the prothoracic segment its protractor and retractor anatomy differ somewhat from that of the meso- and meta-thoracic segments. Details of these deviations and the anatomy of the other thoracic muscles can be found in Marquardt (1940).

The coxa-trochanter joint is moved primarily by the *levator and depressor trochanteris muscles*. The levator is composed of three parts, the anterior and posterior levator trochanteris (both innervated by the C_1 branch of the nervus cruris) and the accessory levator trochanteris (innervated by nl_3). All three originate in the coxa and terminate together at the dorsal rim of the trochanter. The depressor also consists of three parts. One part originates in the coxa (innervated by the C_2 branch of the nervus cruris) and two parts ($L_T tr$ and $L_P tr$) in the thorax (Roth 1980).

The femur-tibia joint is moved by the *extensor and flexor tibiae muscles*. The extensor fibers are relatively short, leading from the dorsal wall of the femur to the tendon which extends practically the whole length of the femur. The flexor fibers are likewise short and originate at the side walls of the femur. In all of the legs the flexor is considerably larger than the extensor.

The extensor tibiae muscle of a middle or hindleg is shown in Fig. 6.3. This muscle contains approximately 150 fiber bundles with a diameter of 0.10–0.15 mm and a length of 1.2–1.7 mm (Bässler and Storrer 1980).

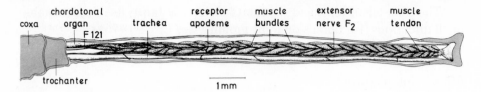

Fig. 6.3. Ventral view of the dorsal half of the femur with the extensor tibiae muscle and the chordotonal organ. The ventral half of the femur has been cut away. (Bässler and Storrer 1980)

The length-tension relationship has been measured on the amputated foreleg (Godden 1974) and on the intact resting (i.e., in the presence of slight SETi activity) and the amputated middle leg (Storrer 1976; Fig. 6.4).

Movement of the tibia-tarsus joint is controlled by three muscles. They make possible the raising and lowering of the tarsus as well as lateral movements and possibly even rotation around the longitudinal axis (C. Walther, pers. comm.).

The tarsus can be bent ventrally by the *retractor unguis*. This muscle has three parts: one which is itself subdivided at the femur base, one in the proximal part of the tibia, and one in the distal portion of the tibia. All three are attached to a single tendon which extends through most of the femur and the tibia to terminate on a highly sclerotized plate, the unguitractor. The retractor unguis muscle has no antagonistic muscle. It works against elastic bands in the tarsus. In the absence of muscle tension the pretarsus is fully extended (C. Walther, pers. comm.).

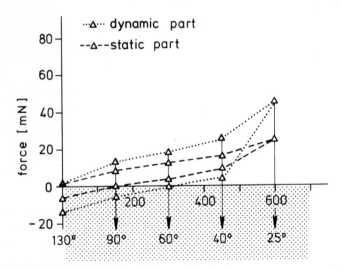

Fig. 6.4. The length-tension diagram for the resting extensor tibiae muscle. The corresponding femur-tibia angles are also shown on the abscissa. (Storrer 1976)

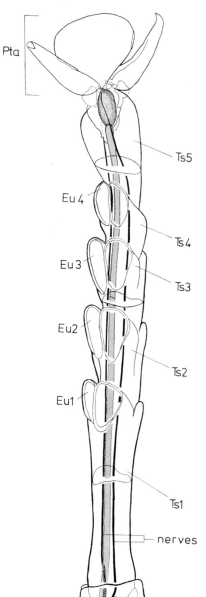

Fig. 6.5. Internal organization of the *Carausius morosus* tarsus. *Pta* pretarsus; *Eu* euplantulae; *Ts* tarsal segment. (Krück 1976)

When the tarsus is resting on the ground, a tug on the retractor unguis tendon bends the pretarsus down. This brings the claws and the gripping pad, the arolium, into contact with the substrate where they exert a distally directed force braced against the gripping pads (euplantulae) of the other tarsal segments and the hairs on the rest of the tarsus which all point distally. The underside of the first tarsal segment has a row of especially sturdy hairs. This arrangement enables the stick insect to cling to smooth as well as rough surfaces.

6.1.3   Nervous System

The meso- and metathoracic ganglia are situated near the leg insertions at the posterior end of their respective segments. The nerves leaving a ganglion and the nerves of a leg are shown schematically in Fig. 6.6. Details of their projections in the body can be found in Marquardt (1940); innervation of retractor and protractor coxae muscles, Graham and

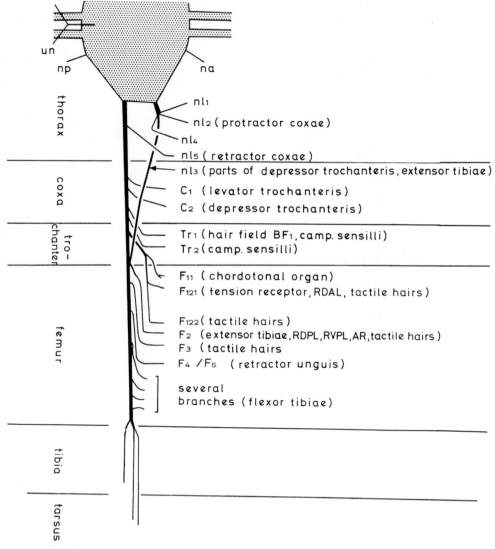

Fig. 6.6. Schematic drawing of the leg nerves including names and primary innervations when known. Thorax and coxa from Marquardt (1940) and Roth (1980); trochanter, Bässler (1977b); femur, Bässler (1977a); tibia, Godden (1972) and C. Walther (pers. comm.); and tarsus, Krück (1976)

Wendler (1981a); coxa, Roth (1980); trochanter, Bässler (1977b); femur, Bässler (1977a); tibia, Godden (1972); and tarsus, Krück (1976).

6.2 Anatomy and Physiology of the Sense Organs of the Legs

6.2.1 Coxa

There are two hair plates on the anterior surface of the subcoxal joint, BF2 near the trochantinal and BF3 near the pleural articulation. BF2 contains 20–30 hairs; BF3 15–20 hairs which are 10–40 μm long (Wendler 1964). On the posterior coxal surface four rows of hairs run approximately perpendicular to the rotational axis of the subcoxal joint. The number of hairs per row lies between 4–5 (innermost row) and 8–13 (outermost row) which are 35–60 μm long (Bässler 1965). Tatar (1976) has described the sensilla on the coxa and trochanter using electronmicroscopy.

Electrophysiological investigations have not yet been performed on the sense hairs. After removal of BF2 (Wendler 1964) or the hair rows (Bässler 1965) an externally applied force causes a much greater deflection of the coxa, suggesting that these sense organs measure the position of the subcoxal joint. The hairs apparently do not adapt completely since continuous stimulation of BF2 produces a distinct behavioral change in the walking animal (Bässler 1977b). No clear function for BF3 has been demonstrated (Wendler 1964).

The coxa also possesses single tactile hairs and campaniform sensilla (Tatar 1976). An internal sense organ which has not yet been identified is mentioned in Section 3.4.

6.2.2 Trochanter

A hair plate consisting of 20–30 hairs is situated near the anterior articulation of the coxa-trochanter joint (Wendler 1964) and is innervated by the Tr1 nerve (Bässler 1977b). Each hair contains one bipolar sense cell (Tatar 1976). These hairs have not been investigated electrophysiologically but behavioral experiments by Wendler (1964) suggest that they have a proprioceptive function.

On the ventral side of the trochanter is a rhomboid-shaped hair plate with about 25 hairs which resemble those of the coxal hair rows (Tatar 1976). The function of this hair plate is unknown.

Two fields of campaniform sensilla are located in pits on the trochanter, one on the anterior and one on the posterior wall. The sensilla in the anterior pit are arranged into two groups of about 12 and 6 sensilla. The longitudinal axes within a group have the same alignment, and the two groups form an angle of 90° (middle and hindlegs) or 110° (forelegs). A third field of campaniform sensilla lies distal to BF1 and is also subdivided into two groups (Tatar 1976). This field and the anterior pit field are innervated by Tr1; the posterior pit field by Tr2 (Bässler 1977b). Electrophysiological recordings have only been made from nerve Tr1 in *Cuniculina*. A few particularly large units can be individually identified in these recordings. They show that the campaniform sensilla respond phasotonically to a certain direction of load on the coxa-trochanter joint (Hofmann, pers. comm.).

The trochanter also possesses solitary tactile hairs of unknown behavioral significance.

6.2.3 Femur

Femoral Chordotonal Organ. This organ was first described by Borchardt (1927). As shown in Fig. 6.7, it lies dorsal at the base of the femur and is connected to the tibia by a cuticular apodeme, the receptor apodeme (Bässler 1965). The chordotonal organ is about 1.5 mm long and consists of two cords which are joined at their base and at the insertion of the receptor apodeme. The dorsal cord contains about 200 scolopidia; and the ventral, up to 40. Each scolopidium has two sense cells whose cell bodies lie in the common base of the two cords (Füller and Ernst 1973; this study also gives an exact electronmicroscopic description of the organ and its innervation).

As yet electrophysiological investigations have only been carried out superficially on *Cuniculina impigra* (see Sect. 2.7.2). Mass recordings from

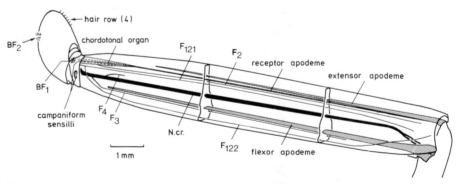

Fig. 6.7. Anterior view of the main sense organs and the innervation of the femur from a middle or hindleg. The transverse sections in the drawing show the three-dimensional nature of the reconstruction

nerve F1 (the common root of the chordotonal organ nerve, F121, and F122) show that the organ contains relatively large, purely phasic units. These may respond to very small amplitude movements. Phaso-tonic units are also present (unpublished). Two findings suggest that the chordotonal organ must also have a tonic component. First, animals whose left chordotonal organs are continuously stretched, veer towards the right when walking on a vertical surface. Insects whose left receptor apodemes have been cut, tend towards the left on the same walking surface. These lateral tendencies persist for several days following the operations (Bässler 1967). Secondly, the femur-tibia control loop has a tonic component (Sects. 2.4.2 and 2.8.2).

Multipolar Sense Cells on the Femur-Tibia Joint. On the dorsal side of the femur-tibia joint adjacent to the soft cuticle are two receptor organs (Fig. 6.8), the RDAL (récepteur dorso-antéro-latéral) with one sense cell and the RDPL (récepteur dorso-postério-latéral) with two sense cells. Both organs respond to joint extension and show continuous low frequency firing only near the extended joint position (Bässler 1977a). A third receptor organ, the RVPL (récepteur ventro-postério-latéral) with two sense cells is located on the joint membrane beneath the flexor-tibia tendon. Its function in *Carausius* is unknown. In *Schistocerca* it responds to a pronounced deformation of the ventral joint membrane (Coillot and Boistel 1969; Heitler and Burrows 1977; Williamson and Burns 1978).

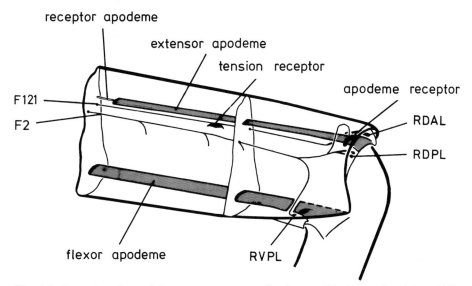

Fig. 6.8. Posterior view of the sense organs on the femur-tibia joint of a right middle or hindleg. The transverse sections show the three-dimensional nature of the reconstruction

Apodeme-Receptor (Fig. 6.2). This sense organ is situated on the ventral surface of the distal end of the extensor tibiae apodeme. It has one sense cell. The nerve leads from the receptor around the apodeme to the dorsal wall of the femur. In *Carausius* no activity could be recorded coming from this receptor. In *Extatosoma* stressing the nerve by lifting the electrode increases its firing rate. In the intact leg the firing frequency reaches its minimum when the leg is in the 90° position, i.e., the nerve is least stretched. Both extension and flexion of the joint increases the apodeme receptor activity (Bässler 1977a).

Tension Receptor. Electrophysiological recordings of nerve F121 reveal a unit that registers the tension of one or a few muscle fibers on the anterior side of the flexor tibiae muscle near its distal end. This receptor probably has a single multipolar sense cell (Bässler 1977a).

Tactile Hairs. Two different types of tactile hairs can be differentiated anatomically. The response of all the hairs to bending is purely phasic and the deflection necessary to elicit a response is quite small. Directional sensitivity could not be demonstrated (Bässler 1977a).

Campaniform Sensilla. A field of campaniform sensilla is located on the proximal end of the femur, ventral to the posterior articulation of the coxa-trochanter joint (Tatar 1976).

6.2.4   Tibia

The sense organs of the tibia have not yet been investigated systematically. There is a subgenual organ. A chordotonal organ is situated on the tibia-tarsus joint and is also connected to the tendon of the retractor unguis muscle by elastic structures. A few campaniform sensilla lie near the femur-tibia joint on the dorsal side (Walther, pers. comm.).

6.2.5   Tarsus

The entire tarsus is densely covered with relatively large tactile hairs except for the arolium and the euplantulae (Krück 1976).

6.3 Motor Innervation of the Leg Muscles

The following description applies to the middle and hindlegs unless otherwise stated. Only the muscles that have been studied in detail are discussed.

6.3.1 Retractor Coxae

The innervation of the individual fibers is similar in the meso- and metathorax. A single fiber in Ra or Rb can be innervated by up to five excitatory axons (SRCx = slow, two or three sFRCx = semifast, and FRCx = fast) and one common inhibitor (CI) (Fig. 6.9). Muscle Rc has similar innervation but only one semifast axon is present. In addition there are also a semifast and a fast neuron which do not innervate Ra and Rb. All axons including a branch of the common inhibitor exit from the ganglion via root nl_5. The slow and inhibitor axon synapses are localized at the tendon end of the muscle fibers. The common inhibitor also innervates the protractor coxae muscles, the extensor tibiae, and probably several other muscles in the leg and thorax (Graham and Wendler 1981a; Igelmund 1980).

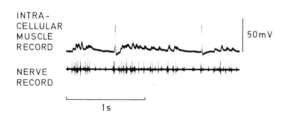

Fig. **6.9.** Simultaneous recordings from nerve nl_5 and a muscle fiber of Ra of the retractor coxae muscle

6.3.2 Depressor Trochanteris

Two excitatory neurons can be distinguished in myogram recordings. One unit appears to be fast, the other slow (Wendler 1972).

6.3.3 Extensor Tibiae (Bässler and Storrer 1980)

This muscle is innervated by three motor neurons: a fast extensor tibiae (FETi), a slow extensor tibiae (SETi), and a common inhibitor (CI) unit which also innervates the protractor and retractor coxae. All three axons exit the ganglion via nerve nl$_3$. For the position of these neurons in the ganglion see Section 2.8.1. The intracellularly recorded responses of these units in the muscle fibers correspond to those of similar units in other insects.

FETi innervates all but the extreme distal end of the muscle. SETi innervates fibers at the extreme proximal end, almost all the fibers in the distal third, and individual fibers in the rest of the muscle. The inhibitor projects only to those fibers innervated by SETi.

If the extensor tibiae muscle is divided into two parts at a point slightly distal to its middle, the proximal portion contains mostly fibers innervated only by FETi and the distal portion, fibers innervated by both FETi and SETi. Such a preparation is useful for investigating the effects of FETi and SETi separately (see Sect. 2.8.2).

Godden (1972) also found two excitatory units in the foreleg corresponding to the FETi and SETi. A third motor unit was found but no intracellular muscle fiber responses could be measured. This unit is probably the common inhibitor.

6.3.4   Flexor Tibiae

The flexor tibiae muscle is innervated by at least two inhibitory units which it shares with the retractor unguis muscle but which are not the common inhibitor described in Sections 6.3.1 and 6.3.3. Both inhibitory units can project to the same muscle fiber but this is rare. Extracellular recordings from the small side branches of the nervus cruris (which project exclusively to this muscle) show that the muscle is innervated by at least 8 and probably 12 or more motor neurons. The innervation of this muscle is quite complicated as in the locust (Theophilidis and Burns 1979). If the muscle is cut into two halves and the chordotonal organ stimulated, the proximal portion shows a rapid rise and decay whereas the distal portion both rises and decays very slowly. Like the extensor tibiae, the flexor tibiae muscle appears to have a fast proximal and a slow distal portion (Debrodt 1980). In the foreleg there appear to be three fast units (Godden 1972).

6.3.5   Retractor Unguis

Two inhibitory units of this muscle have already been mentioned in Section 6.3.4. Their fiber projections were not investigated. In the foreleg the retractor unguis is innervated by seven motor neurons (Godden 1972).

Recordings from the single inhibitory axon could only be made in the proximal tibial end of the muscle. Of the three fast units, two project to the femoral portion and one to the two tibial portions of the muscle. One slow unit innervates all three muscle parts; a second, the femoral and proximal tibial portions and a third, only the proximal tibial portion of the muscle. Walther (1980) found six excitatory and one inhibitory motor neurons in the femoral portion.

References

Autrum H (1950) Die Belichtungspotentiale und das Sehen der Insekten (Untersuchungen an Calliphora und Dixippus). Z Vgl Physiol 32:176–227

Bässler U (1965) Propriorezeptoren am Subcoxal- und Femur-Tibia-Gelenk der Stabheuschrecke Carausius morosus und ihre Rolle bei der Wahrnehmung der Schwerkraftrichtung. Kybernetik 2:168–193

Bässler U (1967) Zur Regelung der Stellung des Femur-Tibia-Gelenkes bei der Stabheuschrecke Carausius morosus in der Ruhe und im Lauf. Kybernetik 4:18–26

Bässler (1971) Zur Bedeutung der Antennen für die Wahrnehmung der Schwerkraftrichtung bei der Stabheuschrecke Carausius morosus. Kybernetik 9:31–34

Bässler U (1972a) Zur Beeinflussung der Bewegungsweise eines Beines von Carausius morosus durch Amputation anderer Beine. Kybernetik 10:110–114

Bässler U (1972b) Der „Kniesehnenreflex" bei Carausius morosus: Übergangsfunktion und Frequenzgang. Kybernetik 11:32–50

Bässler U (1972c) Der Regelkreis des Kniesehnenreflexes bei der Stabheuschrecke Carausius morosus: Reaktionen auf passive Bewegungen der Tibia. Kybernetik 12:8–20

Bässler U (1973) Zur Steuerung aktiver Bewegungen des Femur-Tibia-Gelenkes der Stabheuschrecke Carausius morosus. Kybernetik 13:38–53

Bässler U (1974) Vom femoralen Chordotonalorgan gesteuerte Reaktionen bei der Stabheuschrecke Carausius morosus: Messung der von der Tibia erzeugten Kraft im aktiven und inaktiven Tier. Kybernetik 16:213–226

Bässler U (1975) Zur Definition von Pro- und Metageotaxis bei Insekten. Biol Cybern 19:55–60

Bässler U (1976) Reversal of a reflex to a single motoneuron in the stick insect Carausius morosus. Biol Cybern 24:47–49

Bässler U (1977a) Sense organs in the femur of the stick insect and their relevance to the control of position of the femur-tibia-joint. J Comp Physiol 121:99–113

Bässler U (1977b) Sensory control of leg movement in the stick insect Carausius morosus. Biol Cybern 25:61–72

Bässler U (1979) Interaction of central and peripheral mechanisms during walking in first instar stick insects, Extatosoma tiaratum. Physiol Entomol 4:193–199

Bässler U, Cruse H, Pflüger H-J (1974) Der Regelkreis des Kniesehnenreflexes bei der Stabheuschrecke Carausius morosus. Untersuchungen zur Stabilität des Systems im inaktiven Tier. Kybernetik 15:117–125

Bässler U, Pflüger H-J (1979) The control-system of the femur-tibia-joint of the phasmid Extatosoma tiaratum and the control of rocking. A contribution to the evolution of behaviour. J Comp Physiol 132:209–215

Bässler U, Storrer J (1980) The neural basis of the femur-tibia-control-system in the stick insect Carausius morosus I: Motoneurons of the extensor tibiae muscle. Biol Cybern 38:107–114

Bauers Ch (1953) Der Fixierbereich des Insektenauges. Z Vgl Physiol 34:589–605

Bauchhenss E (1971) Carausius morosus Br. — Stabheuschrecke. Großes Zoologisches Praktikum, Bd 14c. Fischer, Stuttgart

Beier M (1957) In: Bronn's Klassen und Ordnungen des Tierreichs. 5 Band 3, Abteilung 6 Buch 2, Lieferung Orthopteroidea, pp 306–454. Akademische Verlagsgesellschaft Geest u. Portig, Leipzig

Beier M (1968) Phasmida. In: Hdbch d Zoologie, Bd IV, Heft 10. De Gruyter, Berlin

Borchardt E (1927) Beitrag zur heteromorphen Regeneration bei Dixippus morosus. Roux' Arch Entwicklungsmech 100:366–394

Bückmann D (1979) Morphological colour change (2): The effects of total and partial blinding on epidermal ommochrome content in the stick insect, Carausius morosus Br. J Comp Physiol 130:331–336

Buddenbrock W v (1921) Der Rhythmus der Schreitbewegungen der Stabheuschrecke Dyxippus. Biol Zentralbl 41:41–48

Buddenbrock W v (1931) Beiträge zur Lichtkompaß-Orientierung (Menotaxis) der Arthropoden. Z Vgl Physiol 15:597–612

Burger M-L (1972) Der Anteil der propriozeptiven Erregung an der Kurskontrolle bei Arthropoden (Diplopoden und Insekten). Verh Dtsch Zool Ges 65:220–225

Burns MD (1974) Structure and physiology of the locust femoral chordotonal organ. J Insect Physiol 20:1319–1339

Clark JT (1974) Stick and leaf insects. Shurlock, Winchester

Coillot JP, Boistel J (1969) Etude de l'activité électrique propagée de récepteurs a l'étirement de la patte métathoracique du criquet, Schistocerca gregaria. J Insect Physiol 15:1449–1470

Cruse H (1976a) The function of the legs in the free walking stick insect, Carausius morosus. J Comp Physiol 112:235–262

Cruse H (1976b) The control of body position in the stick insect (Carausius morosus) when walking over uneven surfaces. Biol Cybern 24:25–33

Cruse H (1979a) A new model describing the coordination pattern of the legs of a walking stick insect. Biol Cybern 32:107–113

Cruse H (1979b) The control of the anterior extreme position of the hindleg of a walking insect, Carausius morosus. Physiol Entomol 4:121–124

Cruse H (1980a) A quantitative model of walking incorporating central and peripheral influences. I. The control of the individual leg. Biol Cybern 37:131–136

Cruse H (1980b) A quantitative model of walking incorporating central and peripheral influences. II. The connections between the different legs. Biol Cybern 37:137–144

Cruse H, Pflüger H-J (1981) Is the position of the femur-tibia-joint under feedback control in the walking stick insect? II. Electrophysiological recordings. J Exp Biol 92:97–107

Cruse H, Saxler G (1980a) Oscillations of force in the standing legs of a walking insect (Carausius morosus). Biol Cybern 36:159–163

Cruse H, Saxler G (1980b) The coordination of force oscillations and of leg movement in a walking insect (Carausius morosus). Biol Cybern 36:165–171

Cruse H, Storrer J (1977) Open loop analysis of a feedback mechanism controlling the leg position in the stick insect Carausius morosus: Comparison between experiment and simulation. Biol Cybern 25:143–153

Debrodt B (1980) Untersuchungen über die Innervation des Flexor tibiae an Carausius morosus. Diplomarbeit, Kaiserslautern

Delcomyn F, Daley DL (1979) Central excitation of cockroach giant interneurons during walking. J Comp Physiol 130:39–48

Ebner I, Bässler U (1978) Zur Regelung der Stellung des Femur-Tibia-Gelenkes im Mesothorax der Wanderheuschrecke Schistocerca gregaria. Biol Cybern 29:83–96

Eidmann H (1956) Über rhythmische Erscheinungen bei der Stabheuschrecke Carausius morosus Br. Z Vgl Physiol 38:370–390

Ewert JP, Wietersheim A v (1974) Musterauswertung durch Tectum- und Thalamus/Praetectum-Neurone im visuellen System der Kröte Bufo bufo. J Comp Physiol 92:131–148

Foth E-M (1977) Zur Regelung der Stellung des Femur-Tibia-Gelenkes im Mesothorax der Stabheuschrecke Cuniculina impigra. Zulassungsarbeit, Kaiserslautern

Füller H, Ernst A (1973) Die Ultrastruktur der femoralen Chordotonalorgane von Carausius morosus Br. Zool Jahrb Anat 91:574–601

Godden DH (1972) The motor innervation of the leg musculature and motor output during thanatosis in the stick insect Carausius morosus Br. J Comp Physiol 80: 201–225

Godden DH (1973) A re-examination of circadian rhythmicity in Carausius morosus. J Insect Physiol 19:1377–1386

Godden DH (1974) The physiological mechanism of catalepsy in the stick insect Carausius morosus Br. J Comp Physiol 89:251–274

Godden DH, Goldsmith TH (1972) Photoinhibition of arousal in the stick insect Carausius. Z Vgl Physiol 76:135–145

Graham D (1972) A behavioural analysis of the temporal organisation of walking movements in the first instar and adult stick insects (Carausius morosus). J Comp Physiol 81:23–52

Graham D (1977a) The effect of amputation and leg restraint on the free walking co-ordination of the stick insect Carausius morosus. J Comp Physiol 116:91–116

Graham D (1977b) Simulation of a model for the coordination of leg movement in free walking insects. Biol Cybern 26:187–198

Graham D (1978) Unusual step patterns in the free walking grasshopper Neoconocephalus robustus II. J Exp Biol 73:159–172

Graham D (1979a) Effects of circum–oesophageal lesion on the behaviour of the stick insect Carausius morosus. I. Cyclic behaviour patterns. Biol Cybern 32:139–149

Graham D (1979b) Effects of circum–oesophageal lesion on the behaviour of the stick insect Carausius morosus. II. Change in walking co-ordination. Biol Cybern 32: 147–152

Graham D (1981) Walking kinetics of the stick insect using a low-inertia, counterbalanced pair of independent treadwheels. Biol Cybern 40:49–57

Graham D, Bässler U (1981) Effects of afference sign reversal on motor activity in walking stick insects (Carausius morosus). J Exp Biol 91:179–193

Graham D, Cruse H (1981) Coordinated walking of stick insects on a mercury surface. J Exp Biol 92:229–241

Graham D, Wendler G (1981a) The reflex behaviour and innervation of the tergocoxal retractor muscles of the stick insect (Carausius morosus). J Comp Physiol 143: 81–91

Graham D, Wendler G (1981b) Motor output to the protractor and retractor coxae muscles of Carausius morosus during walking on a treadwheel. Physiol Entomol 6: 161–174

Heitler WJ, Burrows M (1977) The locust jump. II. Neural circuits of the motor programme. J Exp Biol 66:221–241

Hoyle G (1975) Identified neurons and the future of neuroethology. J Exp Zool 194: 51–74

Hughes GM (1957) The co-ordination of insect movements. II. The effect of limb amputation and the cutting of commissures in the cockroach (Blatta orientalis). J Exp Biol 34:306–333

Igelmund P (1980) Untersuchungen zur Stellungs- und Bewegungsregelung der Beine der Stabheuschrecke Carausius morosus: Neuronale Grundlagen der Pro- und Retraktion der Coxa. Diplomarbeit, Köln

Jander JP (1978) Zur Kurventechnik laufender Insekten. Verh Dtsch Zool Ges 1978: 258

Jander JP, Wendler G (1978) Zur Steuerung des Kurvenlaufs bei Stabheuschrecken (Carausius morosus). Kybernetik 77:388–392

Jander R, Volk-Heinrichs I (1970) Das strauch-spezifische visuelle Perceptor-System der Stabheuschrecke (Carausius morosus). Z Vgl Physiol 70:125–147

Jörschke H (1914) Die Facettenaugen der Orthopteren und Termiten. Z Zool 111: 153–280

Kalmus H (1937) Photohorotaxis, eine neue Reaktionsart, gefunden an den Eilarven von Dixippus. Z Vgl Physiol 24:644–655

Kalmus H (1938) Tagesperiodisch verlaufende Vorgänge an der Stabheuschrecke Dixippus morosus. Z Vgl Physiol 25:494–508

Kemmerling S, Varju D (1981) Regulation of the body-substrate-distance in the stick insect: Responses to sinusoidal stimulation. Biol Cybern 39:129–137

Kittmann R (1979) Untersuchungen am Regelsystem des Femur-Tibia-Gelenkes der Stabheuschrecke Carausius morosus. Langzeitverhalten des Femur-Tibia-Winkels und der Slow-Extensor-tibiae Aktivität. Diplomarbeit, Kaiserslautern

Krück J (1976) Die Anatomie des Tarsus von Carausius morosus. Zulassungsarbeit, Kaiserslautern

Marquardt F (1940) Beiträge zur Anatomie der Muskulatur und der peripheren Nerven von Carausius (Dixippus) morosus. Zool Jahrb Abt Anat Ont Tiere 66:63–128

Meissner O (1909) Biologische Beobachtungen an der indischen Stabheuschrecke Dixippus morosus. Z Wiss Insektenbiol 5:14–21, 55–61, 87–95

Mittelstaedt H (1978) Kybernetische Analyse von Orientierungsleistungen. In: Kybernetik 1977, Oldenbourg, München, pp 144–195

Pearson KG, Iles JF (1970) Discharge patterns of coxal levator and depressor motoneurones of the cockroach Periplaneta americana. J Exp Biol 52:139–165

Pflüger H-J (1976) Zur Steuerung der Schaukel- und Laufbewegungen bei den Phasmiden Carausius morosus und Extatosoma tiaratum. Dissertation, Kaiserslautern

Pflüger H-J (1977) The control of the rocking movements of the phasmid Carausius morosus Br. J Comp Physiol 120:181–202

Precht H (1942) Das Taxisproblem in der Zoologie. Z Wiss Zool 156:1

Priebasch J (1933) Licht und Farbwechsel bei Dixippus. Z Vgl Physiol 19:453

Rabaud E (1919) L'immobilisation réflexe et l'activité normale des arthropodes. Bull Biol France Belg L3:149

Rahmer C (1972) Zur Regelung der Stellung des Femur-Tibia-Gelenks bei Locusta migratoria. Zulassungsarbeit, Stuttgart

Roth M (1980) Anatomie der Coxa von Carausius morosus. Zulassungsarbeit, Kaiserslautern

Rupprecht R (1971) Bewegungsmimikry bei Carausius morosus (Phasmida). Experientia 27:1437

Schlegtendal A (1934) Beitrag zum Farbensinn der Arthropoden. Z Vgl Physiol 20: 545–581

Schmidt P (1913) Katalepsie der Phasmiden. Biol Zentralbl 33:193–207

Schmitz J (1980) Der „Kniesehnenreflex" der Stabheuschrecke Carausius morosus: Untersuchungen am stehenden Bein eines laufenden Tieres. Zulassungsarbeit, Kaiserslautern

Schmitz W (1978) Die Kontrolle der Körperstellung einer stehenden Stabheuschrecke. Messung der Körperhöhe und der von den Einzelbeinen entwickelten Kräfte. Zulassungsarbeit, Kaiserslautern

Schneider J (1961) Über die Orientierung von Stabheuschrecken (Dixippus morosus Brunner) im Schwereizfeld. Zool Anz 167:93–99

Schreiner HJ (1979) Die Kraft-Höhen-Kennlinie einer stehenden Stabheuschrecke. Ihre Abhängigkeit von der Ausgangsstellung des gesamten Beines sowie vom Beitrag der einzelnen Beingelenke. Zulassungsarbeit, Kaiserslautern

Sharov AG (1971) Phylogeny of the Orthopteroidea. Israel Program Scient Trans, Jerusalem

Stabler RH (1976) The effect of air currents on the behaviour of the Indian stick insect Carausius morosus (Br). Entomol Rec 88:44–46

Stammer W (1978) Die Kontrolle der Körperstellung einer stehenden Stabheuschrecke. Messung der Körperhöhe und der von den Beinpaaren entwickelten Kräfte. Zulassungsarbeit, Kaiserslautern

Steiniger F (1933) Die Erscheinung der Katalepsie bei Stabheuschrecken und Wasser-
    läufern. Z Morphol Oekol Tiere 26:592—708
Steiniger F (1936) Die Biologie der sog. ,,tierischen Hypnose". Ergeb Biol 13:348—
    451
Storrer J (1976) Systemanalytische Untersuchungen am ,,Kniesehnenreflex" der Stab-
    heuschrecke Carausius morosus. Dissertation, Kaiserslautern
Storrer J, Cruse H (1977) Systemanalytische Untersuchung eines aufgeschnittenen
    Regelkreises, der die Beinstellung der Stabheuschrecke Carausius morosus kontrol-
    liert: Kraftmessungen an den Antagonisten Flexor und Extensor tibiae. Biol Cybern
    25:131—143
Tatar G (1976) Mechanische Sinnesorgane an den Beinen der Stabheuschrecke Carau-
    sius morosus. Diplomarbeit, Köln
Theophilidis G, Burns MD (1979) A muscle tension receptor in the locust leg. J Comp
    Physiol 131:247—254
Thomas A (1979) Nervous control of egg progression into the common oviduct and
    genital chamber of the stick insect Carausius morosus. J Insect Physiol 25:811—
    823
Walther Ch (1969) Zum Verhalten des Krallenbeugersystems bei der Stabheuschrecke
    Carausius morosus Br. Z Vgl Physiol 62:421—460
Walther Ch (1980) Small motor axons in orthopteran insects. J Exp Biol 87:99—119
Wendler G (1964) Laufen und Stehen der Stabheuschrecke Carausius morosus: Sinnes-
    borstenfelder in den Beingelenken als Glieder von Regelkreisen. Z Vgl Physiol 48:
    198—250
Wendler G (1965a) Laufkoordination und Regelung der Beinstellung bei der Stabheu-
    schrecke Carausius morosus. Proc XII Int Congr Entomol, London 1964
Wendler G (1965b) Über den Anteil der Antennen an der Schwererezeption der Stab-
    heuschrecken. Z Vgl Physiol 51:60—66
Wendler G (1966) The coordination of walking movements in arthropodes. Symp Soc
    Exp Biol 20:229—249
Wendler G (1968a) Ein Analogmodell der Beinbewegungen eines laufenden Insekts.
    In: Marko H, Färber G (Hrsg) Kybernetik 1968. Oldenbourg, München, S 68—74
Wendler G (1968b) Was messen die Beine von Carausius morosus beim Ermitteln der
    Schwerkraftrichtung? Verh Dtsch Zool Ges 1968:439—444
Wendler G (1972) Körperhaltung bei der Stabheuschrecke: Ihre Beziehung zur
    Schwereorientierung und Mechanismen ihrer Regelung. Verh Dtsch Zool Ges 65:
    214—219
Wendler G (1978) Erzeugung und Kontrolle koordinierter Bewegungen bei Tieren —
    Beispiele an Insekten. Kybernetik 77:11—34
Williamson R, Burns MD (1978) Multiterminal receptors in the locust mesothoracic
    leg. J Insect Physiol 24:661—666
Wilson DM (1966) Insect walking. Annu Rev Entomol 11:103—122
Wilson JA (1979a) The structure and function of serially homologous leg motor neu-
    rons in the locust. I. Anatomy. J Neurobiol 10:41—65
Wilson JA (1979b) The structure and function of serially homologous leg motor neu-
    rons in the locust. II. Physiology. J Neurobiol 10:153—167
Yagi N (1928) Phototropism of Dixippus morosus. J Gen Physiol 11:297
Zwenker C (1973) Die Steuerung der Aktivität bei Carausius morosus. Zulassungs-
    arbeit, Stuttgart

Subject Index

Definitions are given in *italic*

active movement 2, 10, 12, 18f., *65f,* 65–69, 110f.
aiming 116f.
amputation 80f., 120, 126–129, 131, 133–136
Anisomorpha 4
ant mimicry 4
antenna 139ff., 143ff.
anterior extreme position *82,* 82f., 93, 112, 118, 120
apodeme-receptor 158
arolium 153, 158
arousal 63, 65
asymmetry 124, 128f., 131, 135
autotomy 127f., 150

Baculum 45
band-pass filter 39
behavioral physiology 1, 3
behavioral repertoire 2, 6, 7
Bode plot 24f., 27f., 46f.

campaniform sensilla 91f., 105f., 112, 139, 155f., 158
catalepsy 2f., 6, *10ff.,* 10–21, 43–49, 62, 65f.
catch effect 61
center of gravity 78, 141
central oscillator 28–32, *29f.,* 48, 98, 103, 115
central-peripheral program *79f.,* 98, 109, 115
central program *79*
chordotonal organ 13–21, 23–30, 33–62, 65–69, 90f., 96f., 102f., 110ff., 114, 117, 119, 139, 141ff., *156f.*
circadian rhythm 7
claw flexing 6, 64f.
cockroach 78, 103f.
color change 2, 146
color vision 146
common inhibitor 22f., 32, 49, 52f., 72f., 88, 99, 102, 106, 159f.
conduction velocity 56, 59

connective 64, 108
conveyor belt 104f.
coordinating influence 98, 109, 115, 131–136
coordination 78f., 119–136
coordination model 115, 126, 131–136
corner frequency 34, 39, 41, 44, 47f., 54, 57f., 61f.
coxa-trochanter joint 13, 22, 30, 69ff., 89, 114, 150f., 155f.
crossing the receptor apodeme 30, 47f., *90,* 90f., 96f., 112, 114, 117f., 129
Cuniculina 45ff., 49, 56–61, 156
cybernetics 3, 62

death-feigning 2
decerebrate animal 7f., 32, 66ff., 81f., 108, 117, 119, 135
defense clap 4
defense gland 4
denervation 30ff., 99–106, 108, 115
depressor trochanteris 31, 69, 71, 96f., 136, 151, 159
depth of thanatosis 9, 12, 21

egg laying 2
electroretinogram 146
ethology 1, 3
euplantulae 153, 158
evolution 43–49, 51
experimentally induced afference *89f.,* 89–97
Extatosoma 45–51, 95, 118, 158
extensor tibiae 7, 15, 18f., 21ff., 28, 31, 33ff., 37–43, 49, 51, 53ff., 59f., 67, 87, 95, 110, 114, 151f., *159ff.*
eye 6, 145f.

feedback oscillation 27f.
femur-tibia control system 13–21, 23–28, 30, 33–62, 65–69, 110ff.
femur-tibia joint 4–63, 65–69, 86, 90ff., 110ff., 114, 117, 141, 150, 157

Volume 7, Part 6
Comparative Physiology and Evolution of Vision in Invertebrates

A: Invertebrate Photoreceptors
By H. Autrum, M. F. Bennet, B. Diehn, K. Hamdorf, M. Heisenberg, M. Järvilehto, P. Kunze, R. Menzel, W. H. Miller, A. W. Snyder, D. G. Stavenga, M. Yoshida

Editor: **H. Autrum**

1979. 314 figures, 17 tables. XI, 729 pages
ISBN 3-540-08837-7

Contents: Introduction. – Photic Responses and Sensory Transduction in Protists. – Intraocular Filters. – The Physiology of Invertebrate Visual Pigments. – The Physics of Vision in Compound Eyes. – Receptor Potentials in Invertebrate Visual Cells. – Pseudopupils of Compound Eyes. – Apposition and Superposition Eyes. – Spectral Sensitivity and Colour Vision in Invertebrates. – Extraocular Photoreception. – Extraocular Light Receptors and Circadian Rythms. – Genetic Approach to a Visual System. – Author Index. – Subject Index.

B: Invertebrate Visual Centers and Behavior I
By M. F. Land, S. B. Laughlin, D. R. Nässel, N. J. Strausfeld, T. H. Waterman

Editor: **H. Autrum**

1981. 319 figures, 10 tables. X, 635 pages
ISBN 3-540-08703-6

Contents: Neuroarchitecture of Brain Regions that Subserve the Compound Eyes of Crustacea and Insects. – Neural Principles in the Visual System. – Polarization Sensitivity. – Optics and Vision in Invertebrates. – Author Index. – Subject Index.

C: Invertebrate Visual Centers and Behavior II
By H. Autrum, L. J. Goodman, J. B. Messenger, R. Wehner

Editor: **H. Autrum**

1981. 216 figures. IX, 665 pages. ISBN 3-540-10422-4

Contents: Light and Dark Adaptation in Invertebrates. – Comparative Physiology of Vision in Molluscs. – Organization and Physiology of the Insect Dorsal Ocellar System. – Spatial Vision in Arthropods. – Author Index. Subject Index.